THE SACRED COMBE

A Search for Humanity's Heartland

Simon Barnes

Illustrations by Pam Guhrs-Carr

BLOOMSBURY

LONDON · NEW DELHI · NEW YORK · SYDNEY

Bloomsbury Natural History
An imprint of Bloomsbury Publishing Plc

50 Bedford Square   1385 Broadway
London   New York
WC1B 3DP   NY 10018
UK   USA

www.bloomsbury.com

BLOOMSBURY and the Diana logo are trademarks of Bloomsbury Publishing Plc

First published 2016. Paperback edition 2017

British Library Cataloguing-in-Publication Data
A catalogue record for this book is available from the British Library.

Library of Congress Cataloguing-in-Publication data has been applied for.

ISBN (paperback) 978-1-4729-1403-3
ISBN (ebook) 978-1-4729-1404-0

2 4 6 8 10 9 7 5 3 1

Typeset by Deanta Global Publishing Services, Chennai, India
Printed and bound in Great Britain by CPI Group (UK) Ltd, Croydon CR0 4YY

For CLW
who encouraged me to go
and who welcomed me back

1

The Valley

It was the moment when I noticed someone was eating my house that I knew I'd come home. It was my first night in the Valley: perfectly black, the air filled with swishing, ripping, munching. The house in question was a thatched hut a few yards from the river. There wasn't much of a window; the place wasn't designed for anything except sleep, but I got from my bed and peered through the insect-gauzed strip.

There were about half a dozen of them, a cheery and sociable little group lightly snacking on my roof and on the trees that surrounded the hut. Their footsteps made no sound, in the eerie fashion of elephants: vast, bedroom-slippered feet. They were quite unafraid: happy, relaxed, comfortable. I wondered what would happen if I were to step out among them, perhaps to attempt a mystical communion with them, perhaps to run for my life. But I had no real thought of doing either.

It was alarming, but it wasn't frightening. An important distinction. Being about a foot from a group of animals that could tear both me and my hut apart was curiously soothing. There was a thrill of wonder in this — wasn't I an adventurous devil to be in such a place and in such company? But that was only the superficial emotion. Behind it was a great soul-deep happiness: a profound sense of having arrived.

If I could have been sent straight back home — right now, after just four hours in the Valley, back to Hadley Wood, in Hertfordshire, where I lived in those days, a couple of miles beyond the furthest reaches of the Northern Line — I would have been satisfied with the trip and I would have been changed forever. The adventures of the rest of my first trip to the Valley were all wonderful enough, but they were just confirmation of what I learned on that first elephant-haunted night. The world was no longer the same and nor was I. I had been in the magical valley: I had found the sacred combe; I had entered the secret garden.

The great round wet loaves scattered around the hut when I rose at dawn were proof that it was not a dream. Or perhaps it was dream-dung. No matter. I had dreamed all my life of being in such a place; to be there at last, dung, ripped roof and all, blurred the distinction between dream and reality so completely as to make both concepts irrelevant. I walked on into my first day in the Valley. My Valley. Their Valley. Our Valley. Whatever: certainly magical. Sacred. Secret. And above all, home.

2

A Hoopoe

A few years earlier, I read John Fowles's novel *Daniel Martin*: brilliant, but alas, only intermittently. All the same, something of the book stayed with me. A bit like those elephants. The eponymous hero is musing on loss, and with it the sense of finding, or almost finding. He then touches on an author, Nicolas-Edme Restif de la Bretonne, who spent his childhood in a Burgundy village around 1740, and wrote an autobiography, *Monsieur Nicolas*: "He tells how one day, his father finding himself without his usual shepherd, he was allowed to take out the family flock; and how, wandering with it, he came on a secret valley in the hills behind the village. He had never heard anyone speak of it before. It was miraculously lush, green, secret, and full of birds and animals… a hare, a roebuck, a pair of wild boars mating. A hoopoe, the first he had ever seen, flew down and began feeding in a tree of wild honey-pears. He felt these creatures were in some way tamer and more magical here than outside, so that he had a sense of trespassing."

It is something deep in all of us, this valley. It is buried in all our memories, even though we have never been there. It is as old as humanity: this place set apart from the common run, this place where humans are at peace with the rest of creation. We may be trespassers, but the owners don't really mind. Our trespasses are forgiven. The place exists in our imaginations, in our species-memories; it exists in a long-lost golden day of childhood, in the memory of an idyllic doomed love affair; it exists sometimes in an actual place. Often enough, a place we daren't go back to, for fear that it has changed too drastically. Or we have.

Fowles describes it as "a place outside the normal world, intensely private and enclosed, intensely green and fertile, haunted and haunting, dominated by a sense of magic that is also a sense of a mysterious yet profound parity of all existence."

Parity? Parity with elephants? Are we humbling ourselves here? Or exalting ourselves?

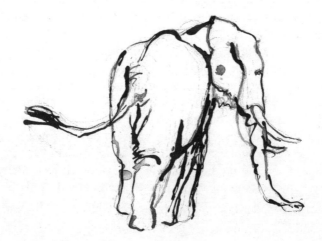

3

Some Famous Elephants

Jumbo
Bought by PT Barnum, eventually joining London Zoo where he acquired his name; *jumbe*, Swahili for chief, is the likely origin.

Kala Nag
Co-star in the Kipling story *Toomai of the Elephants.*

Babar
Green-suited elephant who brought the wisdom of the West to the elephants and became king, in the stories by Jean de Brunhoff.

Castor and Pollux
Inhabitants of the Paris Zoo, eaten by wealthy Parisians during the Siege of Paris in 1870.

Lulu, the Blue Peter elephant
Ran amok in the studio during the television programme previewing the Blue Peter journey to Sri Lanka. My father was a Blue Peter producer at the time.

Hathi
Wise elephant in Kipling's *Jungle Book*, narrator of the parable in *How Fear Came.*

Ganesh
Elephant-headed Hindu god. Remover of obstacles, patron of arts and sciences, master of wisdom, lord of beginnings.

Raja
Carried the eye tooth of the Lord Buddha for many years during the annual Perahera at Kandy in Sri Lanka, walking only on silk, for his feet were too sacred to touch the earth. On two occasions separated by some years, he attempted to attack my father for the various inconveniences he experienced during filming for Blue Peter.

Max
Circus elephant whose death traumatised Pistols Scaramanga in Ian Fleming's *The Man with the Golden Gun*. According to the author of the Secret Service file that M reads: "I see in this dreadful experience a possible reason for the transformation of Scaramanga into the most vicious gunman of recent years".

She–elephant
Vocal during the dawn of creation in Narnia, in CS Lewis's *The Magician's Nephew*.

The elephant's child
From Kipling's *Just So Stories*; he was the best elephant writer. The elephant's child had her short stubby nose stretched into a trunk by a crocodile who wanted to have elephant's child for breakfast.

Toung Taloung
The Sacred White Elephant of Burma, purchased by PT Barnum for $250,000, and ever since an expression used to describe something far more expensive than it's worth.

The Kilimanjaro elephant
Had tusks that weighed 237lb and 225lb; no other tusk has weighed more than 190lb.

Echo

Star of Cynthia Moss's documentary and book *Echo of the Elephants*. "I wanted to have an elephant matriarch that the audience could recognise easily, and I thought she would be the one I could find more easily. She's also very beautiful."

4

The Leopard and
Her Child

I shall never know what happened next. Why did we leave?
I'll always regret it. The Luangwa Valley is famous for leopards
and that first morning, after walking out past the elephant
dung, we set off in a Toyota Land Cruiser and found one. Easy
as that. We were driving past a gathering of baboons, a hectic
village of them foraging and quarrelling and chasing and
teasing and making up: a community full of noise and gossip
and incident. And then a quite different sound: a huge twin-
syllable bark: ba-oo! And another and another: the big males all
standing up tall to utter a dreadful warning. To each other, to
all the tribe, but also to the leopard. No, not leopard but
leopards, for there were two: a mother and cub, the cub three-
quarters grown and male.

How to describe a leopard? I'd better not begin. All accurate
description sounds like excess and to use every excess in the
Thesaurus would be crass understatement. A daylight leopard,
too: a rare thing, for they are creatures of the darkness. Perhaps
they'd had a bad night: perhaps hunger, perhaps the cub's
naïveté had prevented the mother from killing for both the
previous night. Perhaps that's what drew them to visit this
baboon colony by day. Baboons are a favourite food, but in the
daylight the odds were with the baboons.

Fully grown male baboons are fearsome creatures. They
tend to adopt manlike poses, lounging on termite mounds,
forearm resting on knee, fingers limp, as if waiting to take
another pull on a Gauloise. They are strong, armed with
prominent canine teeth. When you find a mammalian social

structure in which one male can have several or many females at once, you tend to find marked differences between the sexes: male lions have manes, male impalas and kudus have horns; female baboons are small and neat while male baboons are not. And as the baboons sighted the leopards the males set off – tentatively and thoughtfully but quite clearly – to confront them. To see them off. The rest of tribe vanished. At this, the leopard parent vanished in a quick slink of gold, but the cub didn't. He decided to climb a tree. Big mistake. Huge.

He'd failed to factor in the idea that baboons can also climb trees. They're better at it than any cat. They have grasping hands and can trust much slimmer branches. And so the male baboons climbed after the young leopard and began to torment him. As we watched, younger and younger baboons joined in and climbed up to shout and swear at the leopard. Every so often he would smash out an elegant spotted paw, and twigs would fly and baboons would leap and whoop, and sometimes the leopard would slip and slither and regain his grasp. It was a stand-off: but with all the advantage to the baboons. We watched for maybe an hour. And then… we went back to camp for breakfast.

Our guide, obeying standing instructions about meal-times, just pulled out and away we went. I thought then, I shall never know what happened. But, I reckoned, obviously this sort of thing happens all the time, so I'll catch this sort of interaction on some future occasion. In all the time I have spent in the Valley, in all the time I have spent elsewhere in Africa, I have never seen anything like it. Did the cub make a break for it, running through his tormenters and suffering blows and bites to rejoin his mother in hiding? Was it a stand-off till night fell and the baboons went to roost, knowing the advantage had shifted? Or did they, like Yahoos surrounding a Houyhnhnm, tear the young leopard apart? Did the mother step in to play a part in a rescue? Or did her instincts tell her that it was better to survive and breed again rather than risk death under those canines?

I shall never know. The Valley, like all sacred combes, is full of
mysteries. But that's a mystery I could have solved.

5

Riding Through
the Glen

I was Robin Hood. But I never did much about it. I just was, and for some reason, that was enough. I had a Ladybird book of Robin Hood: on the cover was a scarlet-hooded man in striped tights, high in a tree launching an arrow at a distant castle. It's a bit of a collector's item now, costs at least ten quid. When I was older I graduated to *The Adventures of Robin Hood* by Roger Lancelyn Green. I never read it all the way through, even though Robin Hood was for years my favourite character. I read two chapters only, and them often: the one where he meets Little John and takes a ducking and the one when he enters an archery contest incognito. The other stories I scarcely or never bothered with. I still don't really understand that: perhaps it was because no text, no matter how authentic, could match the reality of my Robin Hood dreams. Every tale, no matter how well turned, couldn't help but profane, no matter how slightly, my deepest feeling about Robin and the Greenwood.

Then there was the television series, also called *The Adventures of Robin Hood*, the one with the song in it: Robin Hood, Robin Hood, riding though the glen. ("They handled all the trouble on the English country scene, and still found plenty of time to sing…") I enjoyed that of course, but I knew it wasn't real. I understood it then as a spirited but doomed attempt to bring my myth to life. The television series and the books: they could never get it quite right. This was something deeper than acting and words.

It wasn't the stories, though I liked them well enough. It was what lay behind the stories. And what lay behind the stories was the Greenwood. What mattered was the endless landscape of trees, the society of the forest, being an outlaw: leaving civilisation behind to find a deeper and richer life beneath the canopy of oaks. Hollow trees, secret clearings, roads feared by all outsiders, hidden paths though the thickets, aerial routes that only the initiated knew and only the brave could tread: that was my childhood world. That's what being Robin really meant.

It's a very English myth, Robin Hood. We are a nation of hiders, lovers of secrets. The idea of living by stealth and secrecy is deeply attractive to us. We shy away from the open, the up-front, the blatantly obvious. It's what we do, which is why we find a deep liberation when we spend time in more forthright societies like the United States or Australia.

Robin Hood may be a kind of foundation myth for the English, but his appeal goes far beyond England. A Robin Hood website lists more than 40 films and television series, including German, Italian and Japanese Robins. Hollywood has a wrestle with Robin every four or five years: Kevin Costner had Morgan Freeman as the Muslim soldier whose life Robin (that's Costner, of course) saved in the crusades, a man full of exotic mysteries and profound messages about the real nature of brotherhood. Fair enough, nice idea, but Morgan and Kev take second place to the trees. In every western ever made the true star is the landscape. In every Robin Hood film the real star is the Greenwood. And there's a deep yen for the Greenwood, even in Japan.

6

Some other Boyhood Identities

In alphabetical order:

Babar the Elephant
Bagheera (in *The Jungle Book* – and no, not the bloody film)
David Attenborough
Dr Dolittle (not either of the films; the books by Hugh Lofting)
Gerry (in Gerald Durrell's *My Family and Other Animals*)
Hank Marvin (lead guitarist in The Shadows)
Noddy
Peter Pan
Ratty (in *The Wind in the Willows*)
Reepicheep (from *The Chronicles of Narnia* by CS Lewis)
Romany (from the *Romany* books by Bramwell Evens)
Tonto
Toomai (in the Kipling story *Toomai of the Elephants*)

7

An Enchanted Combe

Strictly speaking combes are Devonshire things: the safe and fertile hollows of that lumpy landscape. They are, if you like, valleys only more so: super-valleys in which safety and comfort contrast frightfully with the brutal uplands of the moors. My old friend Ralph, a Devonian for the past 40 years even if he was born elsewhere, tends to get side-tracked on a walk: he likes to include one or two of the high-value Scrabble letters when he walks from A to B. One time he and I got side-tracked into something very like a sacred combe. The valley was so narrow it was almost a gorge: a dramatic cleft in the land, as if God had created it with an irritated blow of a cleaver. No one went there, so far as we could tell; certainly there was no way through. The temperature leapt athletically as we entered, for the wind passed above and couldn't penetrate. It was more humid than the outside world, for the water in the air had nowhere to go. The place was a microclimate: sheltered to an almost sinister degree, an English rainforest fifty yards wide and half a mile long. Common plants grew to heights they would never dare aspire to outside: we had to shoulder our way through a forest of seven-foot docks. It was silent: you could almost hear the trees growing. It had that feeling of interiority that valleys have: an almost indoor version of outdoors. It was safe, in that it knew how to look after itself, but it didn't feel safe for an intruder any more than a strange house would. A pheasant broke cover at my feet, shattering the silence, and I squeaked girlishly, almost throwing myself into Ralph's arms. We carried on in until the place became too overgrown; we scrambled up the sides as far as we could before it got too precipitate and we had to come bum-slithering down. So we turned back. We emerged from the combe as you emerge from a dream: no longer quite sure what all the fuss was about, but ever so slightly out of step with the waking world.

8

The Rift

The Great Rift required a rather bigger cleaver. It is a vast furrow in the earth that runs from Northern Syria down to Mozambique, taking in the River Jordan, the Red Sea and the Dead Sea, the Sea of Galilee, the Wadi Araba and the Gulf of Aqaba before slashing down through Africa across Ethiopia, Kenya and Tanzania. These days it's not considered a single structure but a series of linked geological events. The Great Rift links three of the great world religions – Judaism, Christianity, Islam – with the place where humans first walked upright. It holds the secret of what it means to be human; it also links us humans with the wild world, for unlike most of the rest of the world it still holds big populations of large mammals.

Come down to the southern end of that world-cleaving valley and you find the Mafinga Hills. Here, at about 1500 metres, in Zambia near the border with Tanzania and Malawi, the Great Rift begins a comparatively gentle southwestern extension. In these hills the Luangwa River rises, flowing into the Zambezi 770 kilometres further on. About 300 of these kilometres flow through the North Luangwa National Park and the South Luangwa National Park. And I suspect that the water that flows through this valley has strange properties, and is fully capable of stealing little chunks from human beings who get too close to it. For, you see, I went back. I left a fairly substantial piece of my heart in that valley, and so naturally, I have to go back there every so often. I seem to have no choice in the matter. Those nocturnal elephants, they started something.

9

Love Dawns

There are two breath-taking passages about the dawn of love in Anthony Powell's 12-volume sequence *A Dance to the Music of Time*. The first comes in the third volume, *The Acceptance World*, when Nick, the narrator, is sharing the back seat of a car. When they are "a few hundred yards beyond the spot where the electrically illuminated young lady in a bathing dress dives eternally through the petrol-tainted air", he takes Jean Templar in his arms. "Her response, so sudden and passionate, seemed surprising only a moment or two later. All at once everything was changed. Her body felt at the same time hard and yielding, giving a kind of glow as if live current issued from it. I used to wonder afterwards, whether, in the last resort, of all the time we spent together, however ecstatic, the first moments on the Great West Road were not the best."

Certainly, everything was changed. I knew that from my first few hours in the Valley. I knew even then that I would never see life in quite the same way again. There was a feeling of commitment: certainly a man is more committed to a woman after he has first taken her in his arms. But those first moments, that first night with the snacking elephants, that morning with the tormented leopards and their never-to-be-completed story – however ecstatic – were not the best, nor did I ever seriously imagine them to be. They were just the mad, head-spinning, brain-curdling moments of the first physical contact. So let me move on to the fourth volume of the *Dance*, *At Lady Molly's*. The narrator describes his emotions on being introduced to two sisters: "Would it be too specific, too exaggerated to say that when I set eyes on Isobel Tolland, I knew at once that I should marry her? Something like that is the truth; certainly nearer the truth than merely to record those vague, inchoate sentiments of interest of which I was so immediately conscious." This audacious leap across the entire time-frame of the novel sequence is shocking, against all literary convention. And yet it's something most of us have experienced in one form or another: a sudden cool and absolute certainty about the path that lies before us, an immense decision that for some reason doesn't feel at all like an assertion of the will. It's more like a sudden realisation that an unalterable for-all-time decision has already been made. There is a sense of passivity: as if the decision-making process had nothing whatsoever to do with you.

My first experience of the Valley was like that. Of course, I didn't know I would come back again and again and again. But there was something a good deal more than vague, inchoate sentiments of interest. It was a sense of belonging. It was a sort of proprietary feeling: not one that I had for the Valley, but rather, I felt that the Valley had for me. Whether I came back or not, the Valley had claimed me.

10

Pig in the Middle

It was fear that kept me away from Africa for so many years. Not fear of lions and elephants and snakes. Fear of vehicles. Fear of humanity. How many lion-loving people would I find clicking cameras around the same surrounded beast? (Some years later in Kenya I was to live this nightmare with 30 vehicles surrounding a single serval, a fine, exquisite and quite bewildered little cat.) I wanted Africa to be right. I knew that it might turn out to be something very important; I was terrified of getting it wrong. I read a lot of brochures. I actually booked up one trip and then cancelled it a fortnight later. I got frightened off by a flier the company sent to whet my appetite: "We all waited, not daring to click our cameras…" I didn't want to be among that lot with their bloody cameras. I didn't want to observe and record. I wanted to be a part of it.

I was in mild despair when I read yet another brochure. No, no, no — not that one either. I don't know why I even looked at Zambia. I knew that real Africa was Kenya and Tanzania. Where the Rift flows. I'd never heard of anyone going to Zambia, so I was certain it offered a sub-optimal experience. But I read the Zambian page of the brochure just the same and came across something really rather odd. It wasn't just that they promised lions and leopards. They also offered to take me for a walk. Which is insane. Whatever you do, never leave the vehicle. Even I knew that. And here was a brochure, readily obtainable, apparently from a reputable company, offering the complete madness of walking in the middle of the African bush. How many animals out there could kill you? Walk among them? Are you mad? You bet. Like Corporal Jones in *Dad's Army*, I would like to volunteer to be that man what walks with the lions, sir.

And not very long afterwards, I did. It was walking that changed everything: absolutely everything. I would never have loved the Valley as I did had I seen it only from a vehicle. So I set off, walking with Bob, the guide, and Isaac, the scout. It was like taking a stroll into Eden, into Narnia, into Middle Earth, into Mowgli's jungle, into every magic place that ever haunted you as a child, and whose half-forgotten memory haunts you still. Isaac's job was to wear a green uniform, carry an ancient rifle and make sure that he never needed to use it. This involved a contradiction because he was also required to take us as close to the great beasts as possible. He had rather stately manners, fair English, and a passion for the big mammals. He would borrow a pair of binoculars and gaze his fill, passing them back with a nod – he was not given to easy smiles – and a word. "Beautiful!"

That first walk. Isaac at the front. Bob second; in those days a lot of the guides were white. The group, for this was a group trip, strung out in single file behind. And this was my first experience of that most edgy peace you can find in the Valley: wonderfully calm apart from the fact – or perhaps a little bit because of the fact – that you are walking among creatures that can kill you. By abandoning your vehicle and setting off on foot you became, at least potentially, part of the ecology of the place. I remembered – for I have spent much of my professional life as a sportswriter – Martina Navratilova explaining that the difference between her and her rivals was that they were involved with tennis, while she was committed. So we asked her to explain the difference: "It's like ham and eggs. The chicken's involved. But the pig is committed."

As we set off along the banks of the Luangwa I felt more like a pig than a chicken.

11

Some African Animals that Can Kill You

Lion
Leopard
Hippo
Buffalo
Human
Nile crocodile
Black rhino
White rhino
Puff adder
Black mamba
Elephant
Mosquito
Tsetse

12

A Walk

Walking isn't the best way of getting very close to very large and dangerous animals; if you want that killer photograph, you're better off on a vehicle. But if you want that killer experience, then it's a walk every time.

Speed: you're moving at two or three miles an hour, so your perspective is quite different.

Detail: the bark of trees, the height of a termite mound, nests of weaver-birds, pits dug by ant-lion larvae.

Birds: birding is always better on foot. You see more, and you don't have to stop a vehicle and back up. You're in the same environment as the bird; you're more tuned in.

Sound: without an engine pounding in your ear the whole time you can hear the sharp whistled alarm of the puku, the calls and cries of birds, sometimes the skirring of oxpeckers, which tells you that there's a large mammal just out of sight; the oxpeckers will be feeding on its external parasites.

Flight-distance: you see the antelopes, mostly puku and impala, watching you with their heads up, staring hard, poised for flight but content so long as you come no closer. Those serious eyes.

Scent: with the wind before you, you can get a little closer.

Scent again: sometimes of fresh dung, sometimes of strange vegetation like the bubble-gum tree, and always that tickling spicy sensation in your nostrils that tells you that this is Africa, this is the bush.

Eden: that's always the way it looks when you see that troop of statued antelopes, as if the White Witch of Narnia has just enchanted them: frozen but to a carefully measured extent, trusting.

Paths: they follow the logic of the country, not the needs of humans. Many of the roads are made by hippos on their nightly forays, looking for vegetation to graze and browse. Bush-paths go not from A to B but from food source to food source.

Cosmic courtesy: when you chance upon an elephant, you give way. You're insignificant here.

Sand and dust: you can see what has happened by reading the footprints: the twin sugared almonds of the impala's delicate cloven hoof; the similar but vast prints of giraffe; the clawed pugs of hyena; the thrilling clawless prints of the cats; the great, beautifully veined dinner plates of elephants.

Vulnerability: there are dangerous beasts out here, and besides, you're lost. Could you find your way back to camp if a mad elephant killed everyone but you?

Lion: there's all the difference in the world between seeing a lion from a vehicle and seeing a lion when you're on your own two feet. It's the difference between a pat on the back and a thump in the gut.

First morning in the world. It always feels that way. And you, yourself, new-made.

13

Some Turds

When you walk in the bush you can learn a good deal by looking at turds.

Elephant: like bloody great loaves. Much picked apart by guinea fowl and baboons, because elephant digestion is a pretty coarse mechanism. Wasteful? Not a bit, nothing's wasted here. Look on each elephant dropping as a concentration of resources. Every turd is a convenience-packed seed-bonanza.

Giraffe: impossibly tiny turds from so vast a creature. They really can digest, those boys.

Buffalo: farmyard cowpats.

Zebra: just like you find in a stable.

Impala: little pellets like currants, often piled up in mounds as a territorial advertisement.

Civet: Great Dane sized turds from an animal the size of a spaniel. They too like to make enormous mounds of the stuff to mark territories.

Baboons: unpleasantly like human turds.

Hippo: they chuck the stuff everywhere. When they defecate they fan their tails like a propeller to ensure maximum coverage. It's another territorial thing, but the old story is that God allowed them to live in the river so long as they promised not to eat the fish. So they make a great parade of their droppings to show there's not a trace of fishbone.

Hyena: their turds dry out to pure white; it's the calcium from all the bones they ingest.

Lion: usually somewhat hairy from their rather indiscriminate eating habits.

14

Edenever

Why was this a perfect day? Was it because we were all so happy? Or was it because the place was enchanted? Did the place bring us happiness, or did our happiness bring its magic to the grass, the trees, the bend in the river, the canopy that closed over it? I don't know. I don't really understand the experience at all. I remember a day touched by gold, but I have the haziest memories of what actually happened. I, the oldest of us children, was perhaps eight. I'm pretty sure we travelled in the Bubble: my father's first car, a Heinkel into which all five of us fitted, though it looked as if we were doing so for a bet. My father, then, would have been about 30, my mother a little older. We lived at the time on the top floor of a house in Streatham, South London. This was a family picnic. How on earth did we fit the picnic basket into the Bubble? My father can't remember that detail; my sisters don't remember the day at all.

I don't think we knew where we were going. We headed south, the quickest way out of London from Streatham, and found a place around the Surrey-Kent border. I know that because I asked about the name of the place afterwards. "It hasn't really got a name," my mother explained. "It's just somewhere between Edenbridge and Hever." So I called it Edenever. These days it's a few miles outside the M25. It still counts, I think, as proper countryside. I found that out by looking it up; back then, we just Bubbled along till the buildings got scarcer and the world got greener. Eventually we stopped and wondered if this would do. I think the land was private, I remember crossing a field. And we came to the river and stopped. But why was it magical?

I remember the slow, shallow stream. I have even now a crystal-clear memory of swimming in that river, my face an inch or so above the river bed. I can see myself there now, feel the flow of the water, the exhilarating chill, the sense of oneness with the creatures who lived there. I also know that the memory is entirely false. For a start, I couldn't swim. I am remembering a fantasy, a daydream, a heavenly vision: an experience as vivid as anything I've done in reality. I remember – and this I know is real – the pond-skaters. I don't think I had ever seen them before; I had read about such things, but the experience of meeting them for real was like my first real sighting of elephant and leopard. Perhaps it was a bit like the boyhood experience in *Monsieur Nicolas* and the revelation of the hoopoe feasting on honey-pears – and that must be another false memory, though not mine, because hoopoes don't eat fruit, they take insects and small vertebrates on the ground. I remember water boatmen sculling vigorously on their backs. I also recall a diving beetle – but I think that's another false memory. What I remember most clearly is my sense of a place of wonder and happiness. No doubt the picnic was good: my mother would have made sure of that and the place itself would have made it still tastier. My sisters and I would have giggled and talked and played and had fun together, because we always did, and for that matter still do. My parents must have been truly content that day: a rare moment of peace in hectic and ambitious lives.

You know, I think it was something to do with the roof. The canopy of leaves. The feeling that to get to the river you had to go through a sort of door, a gap in the trees, and once you were there the door closed behind you and the light changed: the direct sun went soft and d\apply and intimate. There was a sense of secrecy about this place: go through the door and a brave new world of brave new creatures was available. It was secret, sacred, cut off, apart from the real world, and yet once you were in, it had more life than anywhere outside. Teeming. Peaceful. Joyful.

We went back there once, perhaps a year later, or a little more, because I remember the field covered in bales of straw. We went there probably at my insistence, but it wasn't the same and we never went back a third time. I managed to find the place again on Google Maps, and learned something useful. That river. I learned its name.

Eden.

15

A Patient Man

The Valley seemed to require a new personality, so I acquired one. There was no conscious effort involved. This was on my second visit. The first had been all open-mouthed wonder. I could scarcely identify a bird, since almost all the birds were new to me and I was too entranced to learn them. I didn't really try very hard: you don't, after all, wonder too much about dates and architectural styles and ownership and history when you make your first trip along the Grand Canal in Venice. You just gawp. And I, for a week, gawped. I was almost silent: a rare thing for me. I remember the first time I knowingly met an unequivocally gorgeous girl. I was perhaps ten at the time; she was about 16. Speech of any kind was wholly impossible both during and for some time after.

But I went back to the Valley: the first of many occasions when I travelled as a writer, and sometimes also as co-leader of a trip. And I found that if I was to forge a relationship with this place beyond the silent gawp – a legitimate response but one that will only get you so far – I needed to find two new traits of personality. And it was easy.

Few people know me as a patient person, still less a serene one. Stand with me on a train platform and I will drive you mad: insisting on arriving hours before departure, staring at the information boards as if they were at any moment to announce the winner, suffering the agonies of the damned at any hint of delay. Sit beside me at a sporting event in the days when I was writing for *The Times* and you will observe frenzy as deadline approaches. The smallest technical malfunction is a disaster on the scale of the assassination of Grand Duke Franz-Ferdinand.

And yet on that trip, staying at Nsefu in the heart of the park, I was commended for my patience. I was surprised, obviously, and wondered how on earth they'd got that impression. And then I realised that I had spent quite a lot of time sitting very still and quiet, just looking. There was a hide with a good view of trees and a small water-hole where birds would often come quite close, or I could just sit by the river and see what happened to pass by. Occasionally checking things out with field guides and reference books, but mostly waiting to see what would turn up because the process was, in a quiet way, enthralling. It wasn't bird-spotting, neither in the sense of ticking things off a list nor in the sense of analysing the species-mix. It was more about beginning to come to an understanding of the place itself. Leave me, come back an hour later, and find me still there, still doing the same not very much. The sense of peace, serenity, oneness, was pleasant enough, but I never sought such things. That wasn't why I went there, and I don't suppose it's why I keep coming back. Serenity was just one more thing in the Valley's gift. I went there to find the Valley, not to find serenity, still less to find myself. But it happened that serenity was the key, or one of them, so I acquired it with some ease and used it to open the door that led me far deeper into the Valley than I would have dared to presume on the night of the snacking elephants and on the morning of the troubled leopards.

I left Nsefu and moved on to stay with Iain McDonald in North Luangwa National Park: to spend a week walking, camping out at night. This wasn't a tourist safari for guests: it was an exploration, preparing the way for Iain's commercial operations later that season. So we walked hard and long and far, and we didn't go through the rewarding wildlife-thronged riverine sections of the park, we climbed over ridges through the places where large mammals seldom come. And I found there not only endurance but a strange content in the monotony of it: just being in the Valley was enough. A year or so earlier I would have looked on such a trip as an ordeal to be avoided at all costs; now I was prepared to walk for ever across a countryside that seemed to have no end, and its endlessness was precisely the point. And there was another thing I had to learn as well: it was a lesson about fear. It was a lesson that required another brand-new personality trait, and that too was easy. But I'll come to that in a moment.

16

Ask Alice

The whole point of paradise is that it's lost. There wouldn't be any point in a Paradise Found. Paradise is not a present joy, it's an appalling nostalgic ache, an eternal rebuke, a constant reminder of failure: failure as an individual, failure as a species. It's also a promise of future bliss, forever just out of reach. We can secularise our culture all we like, but we still yearn for an elusive paradise: one we almost had but lost, one we might have some day if only things could be slightly different from the way they are. If we don't believe in an after-life we still cherish the idea of some elusive perfection: one that in the unimaginative is associated with Caribbean beaches and a drink with an umbrella in it, a timeless vista that will never pall. I know a man – most of us do – who traced his first love on the internet and they met up and a little while later they ran away together, both breaking working marriages to do so. Sometime further on he was asked how it was all going. "Remember why I left her in the first place…"

The first meaning of Eden is our expulsion. The second meaning is that it's now forever out of our reach, forever just beyond our scope. Here's *Alice in Wonderland*:

"Alice opened the door and found that it led into a small passage, not much larger than a rat-hole: she knelt down and looked along the passage into the loveliest garden you ever saw. How she longed to get out of that dark hall, and wander about among those beds of bright flowers and those cool fountains, but she could not even get her head though the doorway; 'and even if my head would go through,' thought poor Alice, 'it would be of very little use without my shoulders. Oh, how I wish I could shut up like a telescope! I think I could, if I only knew how to begin.'"

Of course, she does manage to shut up like a telescope, but by then the key is back on the glass table and out of reach, so she yearns to grow big enough to reach it. And of course she does so, but carries on growing until she is nine feet tall: "Poor Alice! It was as much as she could do, lying down on one side, to look through into the garden with one eye; but to go through was more hopeless than ever: she sat down and began to cry again."

We believe childhood should be an idyll; Alice's predicament with the garden can be understood as a grownup's nostalgia for childhood: a Freudian would go further and interpret it as the desire to escape the traumas and responsibilities of sexuality. No matter which way you slice it though, this is a parable about the unattainable paradise; a paradise that matters not just because of its beauty and perfection but because it is lost. Loss is what matters: loss is what defines a paradise: without loss, no paradise can exist. Or so it seems. We grow up, and leave the idyllic childhood behind. We grow older still and that first fumbling innocent love affair has something that the daily routines of marriage can never possess: a profound beauty that can only come from loss.

But sometimes we can find a glimpse of an unlost paradise. It can come in the discovery of an enchanted combe, and if ever that extraordinary thing happens, it is as if we had cheated our entire culture and civilisation. Such places, such experiences exist in three distinct ways. We can find them in our nostalgia for what was lost, we can find them in a forlorn hope that they can be found again – but sometimes, though rarely, rarely, we can find them the hot light of the eternal present. So share a beer with me: a Mosi of course, the excellent beer of Zambia. Let me tell you about the finest beer of my life, and the competition for that accolade is pretty intense.

It was my first trip to Africa. I was travelling with a group and we had just driven hair-raisingly close to a pride of lions on a kill, so close I could have leapt off the vehicle and rolled about with them, and for that matter, was sorely tempted to. They were the first lions of my life, and I was not about to take them lightly. After some considerable time among them we pulled back, and as the sun came down we stopped for the traditional drinks. Sundowners. Always, from that day to this, a Mosi for me. I sat on the river edge, a typical Luangwa cliff, dropping sheer for 20 feet to the dry river bed: the actual stream was a hundred yards off, for this was the peak of the dry season. I dangled my legs over the edge and leant against a tree. About 200 yards off, a lioness had adopted roughly the same pose, her paws dangling over the edge. She had eaten her fill and was resting, gazing out at the eternal Luangwa. Fifty yards beyond her, another lioness, paws equally a-dangle. And fifty yards again, yet another. All was peace. I raised my bottle and drank Life.

17

A Brave Man

The bank of the Mwaleshi River in North Luangwa National Park is famous as the place where a man was taken from his tent by a lion and devoured. You can imagine, then, how I felt in a tent on the banks of the Mwaleshi River when entirely surrounded by lions. Their music was everywhere: the ground shook with it. A real lion roar is not the petulant snarl you get at the start of a Metro-Goldwyn-Meyer film: it's a long-distance communication device. It comes in a series of great booming resonant bellows, a crescendo that reaches a point of perfect intensity and then falls away. You can hear it from more than a mile off: much further on a still night like this. It works particularly well along water-courses: and lions know this and love to exploit it. They knew we were there, and for these lions we were a novelty: few people make it to the north and no one had been here for at least six months. And so they sang: perhaps to threaten us. I wondered – but some years later – if it wasn't also to reassure each other. So how did I feel?

Wonderful, of course. Thrilled by my own daring, thrilled by the lions, thrilled to be in the heart of this most wonderful of all valleys. I felt quite gloriously brave... and yet, as I lay there in the dark, I pondered my courage and soon realised that it was no courage at all. It is not really so terribly brave to do something you want to do. It's brave to dive into the river to save a drowning child; it's brave to stand up before the world for what you believe in; it's brave to do the right thing when the wrong thing is so much easier – but it's not brave to seek and to find joy. I knew fear that night, of course I did. But I didn't master fear because I was brave; I mastered it because I loved being there among the lions within the walls of the great Valley. I have done some wild and crazy things with horses, but I was never brave: I loved (and love) the whole horsey thing far too much for that to be a factor. That night on the Mwaleshi didn't tell me that I was braver than I thought. I didn't discover courage: I discovered that my tastes in love were wilder than I had ever imagined. If I had a lion's voice I would have trumpeted that love the length and breadth of the Mwaleshi and of the Luangwa River into which the Mwaleshi flows. As it was, I lay there in the dark, part of me lost in the moment of tumultuous wildness and another part of me wondering how and when I could get back to the Valley, to my enchanted combe.

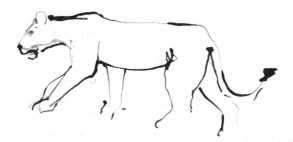

18

Mammals Sighted in North Luangwa National Park

…because for once I kept a list

Chacma baboon
Vervet monkey
Mopane squirrel
Spotted hyena
Leopard
Lion
Burchell's zebra
Warthog
Hippopotamus
Blue wildebeest
Grey duiker
Sharpe's grysbok
Impala
Buffalo
Greater kudu
Eland
Waterbuck
Puku

Paradise within Paradise

The jungle of the Seeonee Hills was another of the enchanted places of my boyhood. It was where Mowgli lived, sharing his life with Bagheera, the black panther who called him brother; with his tutor, the bear Baloo; with Kaa, the giant python; and with the wolves of the Seeonee wolf pack. My heart dwelt in the land where Chil the kite brought home the night that Mang the bat set free.

The boy who was at home in the jungle, who talked to the animals in their own language, who could climb almost as well as he could swim, and swim almost as well as he could run, who was rescued from the Bandar-log, who won the war with the wounded tiger and grew up to be master of the jungle... this was perhaps the foundation myth of my boyhood.

Kipling was a writer of subtleties and contradictions. The jungle was a place of fierce beauty and innocence and perfection: it was my paradise, the more meaningful for being forever beyond me. But within the jungle the creatures have a mythology of their own: a story of a paradise they too had lost. The story is told by Hathi the elephant in *How Fear Came*, and he tells of a time before Fear. "In the beginning of the Jungle, and none know when that was, we of the Jungle walked together, having no fear of one another. In those days there was no drought, and leaves and flowers and fruit grew on the same tree, and we ate nothing at all except leaves and flowers and grass and fruit and bark."

This then, is a paradise within a paradise: the enchanted combe that lies within the enchanted combe. It's a vision of peace that finds many echoes elsewhere: "The wolf also shall dwell with the lamb, and the leopard shall lie down with the kid; and the calf and the young lion and the fatling together; and a little child shall lead them," as it says in Isaiah.

But I have never wanted the Luangwa Valley to be like that. I treasure it for its dangers, for its ferocity, for its carnivores. Once, while I was looking for a gorgeous and eternally elusive bird called the Angola pitta, I walked into a herd of elephants: the best and most vivid kind of birding. This is a place where I've seen hyenas butchering an impala, where I have seen lions slaying a helpless buffalo calf, a place where I have had my share of scrapes. I don't want it to be a gentle land. I don't want it to be safe. You can be safe almost anywhere these days. Perhaps the moral here is that no paradise is or should ever be safe.

20

Rio By the Sea-oh

I have never been able to make anybody understand the true pathos of the situation, and I don't suppose I ever will. I had five days alone in Rio on full expenses with a smart hotel to stay in and the experience filled me with grey horrors of desolation. When I try to explain, people envy my luck and wonder what the hell I got up to. I got up to nothing. I walked a lot; I read a lot in my hotel room; I slept alone in my hotel bed. And, er, that's it.

It was a strange business. The year was 1990. I had been covering Ayrton Senna and the Formula One stuff in Sao Paolo, and was going on to the Test match cricket in Barbados and Antigua, so I was stuck between stories – all this as part of my sportswriting duties for *The Times*. And once I got to Rio I found myself utterly lost and horribly alone. Please don't judge me too harshly. I had planned to get out at once to some wild place and do a spot of wildlifing, but alas, they changed the currency on me. It followed, then, that no one would exchange my good dollars for Brazilian money. Everything was frozen. It was one of many attempts to sort out the eccentricities of Brazilian financial life – famous newspaper headline: *Our Football is Like our Inflation! 100%!* – and I was caught between currencies. This was, for those interested, the launch of the third *cruzeiro*, a currency that was to last for three years. The hotel was happy to take my credit card and reckon the bill in dollars, so there was no danger of either hunger or thirst.

But no local travel company would do the same thing. I managed to get a little drinking money out of the hotel cashiers. I couldn't afford anything as exotic as food, so it was back to the ghastly old hotel buffet every night. I didn't know so much as a cat in that town and even if I'd had loadsamoney, I'm not the sort that walks into a bar and instantly has a dozen best friends. Nor have I acquired a taste for professional sex. I reckon I'm a pretty good traveller, and reasonably self-reliant, always prepared to make the best of what I find, but what with one thing and another, especially the lack of money in my pocket, I was unable to engage with Rio. I spent my time there wishing I was somewhere else: not an experience I've often had. I was pretty well stuck: banged up in five-star comfort, walking by day and reading by night.

What I read was a book about Africa: Peter Mathiessen's *The Tree Where Man Was Born*. I tried to re-read it a few years later: it's all right, but it's not great. But back in Rio it was the Bible, Shakespeare and the *I Ching* all in one: a book that contained the one great truth about where and perhaps even who I needed to be.

"Of all African animals, the elephant is the most difficult for man to live with, yet its passing – if this must come – seems the most tragic of all. I can watch elephants (and elephants alone) for hours at a time, for sooner or later the elephant will do something very strange such as mow grass with its toenails or draw the tusks from the rotted carcass of another elephant and carry them off into the bush. There is mystery behind that masked grey visage, and ancient life force, delicate and mighty, awesome and enchanted, commanding the silence ordinarily reserved for mountain peaks, great fires, and the sea."

The book was a finger pointing to the Promised Land: a paradise out there to be regained. It was a way out of Rio, a way out of every city. I was in Rio and all I wanted was to be in Africa. To be in the Valley: a place I had visited just the once. I had travelled 6,000 miles in the wrong direction to learn that the Valley already had a mighty hold on me. In my hotel room with a belly full of hotel beer I vowed that whatever else I did in my life, I would one day make an extended stay in the Luangwa Valley. There's a Fred Astaire film – his first with

Ginger Rogers as a matter of fact, though neither plays a leading role – called *Flying Down to Rio*. Fred sings:

> *Hey, fella*
> *Twirl that ol' propeller*
> *Got to get to Rio and we've*
> *Got to make time*

Not me, Fred. Got to get back to that ol' Valley. I could hear it calling. Give me a lifelong season ticket to the Mardi Gras in Rio, I'd swap it for ten minutes in the Valley and count myself a pretty ruthless man when it comes to a bargain.

21

Big Boy and Little Boy

The year was 1997 and the telephone was red-hot. Old Luangwa Hands were ringing me up and urging me into action. Norman was dead: what was I going to do about it? So I rang *The Times* obituaries and laid out my case: yes, they said, sure, he's certainly worth a piece, could I get a photograph? It was, it seemed, a quiet day for death, so Norman would get a good show – if only I could find a decent photograph. So I rang another OLH who lived in Cornwall and asked him: "Have you got a copy of that picture of Norman with Big Boy and Little Boy?" He thought so, yes. "If you can get it to *The Times* we should have the lead obit." So off he went to *The Cornishman* newspaper, where he had a mate. As a result the picture was wired – yes, that's what life was like back in those days – to *The Times* and we got the page-top all right.

How could we not? The picture showed a small wiry fellow in enormous shorts and khaki bush-shirt. What rescued the picture from the humdrum were the two immense black-maned lions. He had them both on a lead. Big Boy and Little Boy: later released into the North Park, where they were seen a year later – so they survived as wild beasts after being hand-reared by Norman.

Norman Carr made the Valley possible. He was born in Chinde in Mozambique in 1912, went to school in England, returned to Africa and was appointed elephant control officer in the Luangwa Valley. Between 1940 and 1944 he served in the King's Rifles in the Abyssinian Campaign, leaving as captain. He then became one of the first game rangers in Northern Rhodesia, back in the Luangwa Valley. In 1949 he did a deal with Senior Chief Nsefu, under which income from camps, in what's now called the Nsefu Sector of South Luangwa National Park, went to the local community: a revolutionary notion. Nsefu

Camp was the first tourist camp open to the public in Northern Rhodesia. Norman spent five years as a professional hunter, retiring after he was injured by a buffalo. He then went to work on the other side of the country in the Kafue National Park. He returned to the Valley in 1960 and retired from government service to work in tourism. Like many others he turned his back on hunting and became a conservationist. He did so by setting up the tourism industry: something that became increasingly important as the country became independent. In 1964 they dropped the Northern Rhodesia bit and became Zambia.

Norman established camps and trained guides. You remember that bit about never leaving the vehicle, even for an instant? Norman pioneered walking safaris: seeing the bush on foot, so that you cease to be an observer and become a participant. The Luangwa Valley exists as a thriving place for humans and other animals because of Norman.

I knew him, never well. We shared the occasional beer, the occasional meal when I stayed at Kapani, headquarters of what is now Norman Carr Safaris. He would talk dryly about the bush and its anomalies, never with any self-advertisement. He was no showy-offy hard man of the bush: he just loved the place. Was potty about it. And he had no nostalgia for the time when it was just him and the bush and his great friend and helper, the scout who had the rather unexpected name of Rice Time. He had the generosity to share his enchanted combe with the rest of the world: something that would make better, wiser, richer people of the visitors and safeguard the beloved Valley. He loved the fact that the Valley was a happening place: a place that was valued, a place that wasn't going to be destroyed. For very many reasons, the place now matters too much.

His grave is in a stunning ebony glade close to the Luangwa River. It reads:

Norman Carr
Conservation MBE
May he rest in peace here in the quiet of the park which will
forever be his monument.

There is an inscription to Sir Christopher Wren in St Paul's
Cathedral: *Si monumentum requiris, circumspice*: if you require a
monument, look about. It strikes me that Norman's monument
is better. A cathedral is a fine thing for humans to create, but an
enchanted combe is eternity.

22

Mchenja

Mchenja is ebony. Mchenja bush camp stood – and still stands – in the heart of the South Luangwa National Park, by a glade of these perfect trees on the banks of the Luangwa. There had been some reckless innovations since my last visit: the place was now positively sybaritic. The huts – thatched laundry baskets – were no longer floored by the sand of the river. These days you stood on concrete. Look-shurry. I wasn't sure I approved, but I was desperately glad to be there. I had a hut to myself; it was to be my home for the next two months. I felt, it's true, some slight sense of trepidation. Would I get on with Bob in such proximity for such a time? How would the other staff of the camp deal with my presence? Jess, the caterer, later told me that she feared the worst: a friend of Bob's! For two months!

I hoped that my marriage would survive this absence, and that my professional life would be able to put itself together again after this reckless unilateral sabbatical. I was also concerned about my love for the Valley. Would it really be enough to last me through two months of absolutely nothing else whatsoever? This was only my third time, after all. I had made that first world-changing visit, and the second visit in which I learned something about patience at Nsefu and learned something about courage in the North Park. But I was assuming rather a lot. I was assuming that I was at least in part a person of the Valley, rather than a visitor. But I didn't know for sure.

Bob met me at Mfuwe airport. Bob, guide for part of my first trip to the Valley, had stayed in touch. He was a great eccentric, and a greater birder, and we had fixed up this trip together, with the help of a few others. He was driving a smart new Land Cruiser and we then set off to look for Derek, to take him back to camp. This involved a mazy journey through the little villages, greeted avidly by children and eyed impassively by women.

"Where's Derek?"

"I do not know."

Eventually we found someone who knew where he was. "He is fine, he has been acquitted." So then we found Derek.

"No. It is my day off."

"We need you back at the camp."

"I have things to do."

"What things?"

"Chopping wood."

"Can't someone else do it?"

"No."

So we went back without Derek.

"There is a snake in the showers," Aubrey told me when I arrived.

The showers were carved from living river-bank, with the fourth wall the air of the Luangwa Valley, so you could watch hippos and keep a very decent bird list while washing. I remembered Aubrey from before and the same confident, can-do attitude. "Don't worry, it is harmless."

It was dusk. The clients were on a drive with Manny, so Bob and I had a beer and began to renew acquaintance. It's all too easy to get the wrong idea of Bob. The stories of his eccentricities and his disasters lead you to expect some sort of maniacal clown, but Bob was also profoundly intelligent and in his own way, deeply civilised, with a rich and subtle sense of humour. So we set off on a quick night-drive of our own. I handled the spotlight and to my delight I found a white-tailed mongoose, two genets huddled together in the crotch of a tree, and a giant eagle owl. I have all these details from my notes.

We got back to camp and I met Manny, who was speaking to his clients in exemplary Italian. Manny was a recently qualified guide, one of the first black Zambians to lead walking safaris in the Valley. He was now in his second season as a fully qualified walking guide: young, smart and deeply comfortable in himself. He had found our guests four leopards, a pretty good haul even for the Valley, so they were all as high as martial eagles. I also met Jess, who was aged 23 and had arrived in Zambia after six months overlanding across Africa.

That night I lay on my bed listening hard. There was the constant tinkling of the frogs, the rhythm section of the crickets. I could hear the proop of African scops owl, an easy one, that. Occasionally the whoop of hyena. And then the distant thunder of lion. And already I knew that two months wouldn't be enough, but it was what I had, and so I embraced the coming weeks with joy.

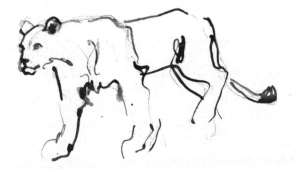

Book-list

Here are the books I took for two months at Mchenja Camp.

Robert's Birds of Southern Africa by Gordon L. Maclean
Portraits in the Wild by Cynthia Moss
Ulysses by James Joyce
Collected Poems of Gerard Manley Hopkins
Three foolscap notebooks

Some Bob Stories

Bob falls from a tree

He had climbed a tree by an established – if primitive – ladder, to investigate a raptor's nest. He fell off, or a rung broke, and he was knocked cold. He spent the night out in the open. The following morning he returned unharmed to Kapani lodge, dripping with blood and asking politely for a beer.

Status: mostly true but exaggerated in places.

Bob drives off the pontoon

He was making a crossing of the Luangwa by the pontoon – a moving cabled platform – when his vehicle came off into the water. It happened because he was looking at a small brown bird.

Status: certainly the vehicle entered into the water, but Bob always vigorously denied culpability. He said the pontoon-man failed to attach the vehicle properly and there was no small brown bird.

Bob drives into the Bangweulu Swamp

He was visiting the swamp in search of shoebill – a bird that's a cross between a heron and a pterodactyl – when he missed the causeway that runs below the surface of the water. The villagers dismantled a hut and used the king-pole to lever the vehicle back onto the causeway, a process that took the entire day.

Status: I know that one's true, I have a witness.

Bob falls from the riverbank

It was Jess's first night at Mchenja; Bob was pointing out the stars to her in a wondrously clear sky. In his enthusiasm he stepped off the bank and landed on his back in reasonably soft

mud about ten feet below. It was a long moment before he
moved: "I've been here six hours and I've killed him already,"
Jess thought.

Status: Jess's word on this.

Bob walks into a lion

Bob walked straight into an angry male lion that leapt out of a
bush and menaced him before running away.

Status: I was there. In fact I was in front of Bob at the time.

25

Cross Lady

It was a quiet morning. Just me, Manny, Perry and a toughish, well-travelled English lady. Perry was the scout, a tall and companionable man. It had been a nice walk, but then it always was. We found ourselves at the edge of a small ebony glade with a small group of elephants: half a dozen of them, none very young, all female so far as I remember, though there may have been a young bull who had not yet begun to make his own way in the world. We stopped and watched them for a while. This wasn't a process of scientific observation, nor was anyone taking photographs. We weren't observing, we weren't spying, we were just – well, being with them. Sharing the same air. Distance: say 50 yards. Not far off, but still respectful. You could hear the swish and munch: they were eating living branches, chopping them off at a convenient point as if they were strands of spaghetti, or picking up a clump of grass and dusting it off against an ankle with whippy movements of the trunk. They knew we were there, and were happy to carry on eating. If they're eating, they're content: one of the founding principles of bush-walking. So all was still, all was balanced; all was a perfect stasis, observers and observed both understanding the rules. The butt of Perry's rifle was on the floor, he had one ankle crossed over another and was using the gun as a leaning post.

And then all of a sudden, one of the younger females got fed up with us. She turned away from the food and made a great lumbering run at us, taking the long Groucho-Marx strides that hurrying elephants go in for. She wasn't that big, as elephants go, but she made herself big enough: ears spread, those two big great maps of Africa filling our eyes. And at around seven foot high at the shoulder, she was at least a respectable class of animal.

Neither Perry nor Manny moved a muscle. Then, when the elephant was 15 yards off, they both clapped, Perry nonchalantly uncrossing an ankle and resting the tip of his rifle-barrel against a hip to do so. At this hardly threatening behaviour the elephant performed a hand-brake turn and went back to the herd, as if she had just committed a rather embarrassing faux pas. We continued our walk. It was still a lovely day.

26

The Lion Dream

Never a day goes past without thinking of Africa. Without thinking of the Luangwa Valley. It's not a permanent longing to be somewhere else: it's more that the Valley is part of who I am, how I think, how I face the world. I often dream about the Valley, too. But more often than not – far more often than not – it's a bad dream. The land of my dreams is also the land of my nightmares. Usually I'm on my own, out there in the bush and always on foot. I somehow know that there's no vehicle for miles. I have to get back somewhere. At first all goes well. I'm generally in open country and I can see for miles. I can see zebras, impalas, pukus, not well, but as part of the restless haze of heat and grass. I'm sort of comfortable, sort of easy with everything. But slowly, like a drop of icy water trickling down my spine, the mood changes. Gradually there is a sense of threat. And just as gradually the threat begins to turn solid. That's when I become aware of the lions. They're watching the zebras and the antelopes. But slowly it becomes clear that they are also watching me. Sometimes they're walking through the grass with that slouching swagger that more or less defines a lion. I know I must walk away from them, but I also know that I must do this without actually running away. I must make a big circle in a manner that won't attract their attention. For some reason that doesn't seem to be possible. Tension is now building up in a horrible and endless silence. I know I am utterly outmatched, utterly out of my depth. I have no idea what to do. And I know that's not the case with the lions.

And then, I'm happy to say, I wake up. Extremely reluctant to attempt sleep for a while. It's a dream based on a real experience: the walk (with Bob, of course, who else?) in which a black-maned lion appeared from nowhere to confront us at a distance of 20 yards. I can see that lion whenever I choose, and quite often when I don't.

So let me tell you about two bad nights. First one: I was 20, my heart newly-broken, living in the absurd hope that she would change her mind, see sense, realise the error of her ways and come back. One night, at the peak of all this, she did. Something woke me, or startled me out of my reverie of sadness and self-pity. A noise in the street. I looked out of the window and, God be praised and forever thanked, she was there. I could see her quite clearly, though only for an instant, looking a little thoughtful, preoccupied, aware of the immense significance of what she was doing. She was dressed as ever with startling style and originality. She was beautiful. She was coming back to me. I waited in joy for the sound of the big door closing below, and the soft sound of her steps on the stair. I waited for the scratch of the key in the lock. And I never heard it at all. Because she wasn't there.

Second bad night: maybe 20 years later. I was in a hut on the edge of the Luangwa River. The entire pride had been roaring with dreadful proximity for some time. I looked from the window − if you can dignify the gap in the grass wall with such a name − and saw that same black-maned lion walking − slouching swagger − past my hut in the shadows of the ebony trees. I tasted, savoured the fear. And I looked the next morning for the great paw prints. They weren't there.

Those are the sorts of things love does to people.

A Fool in Wales

But I've done much stupider things with Bob than walking into lions, and come closer to death, too. There was, at least arguably, an element of bad luck about the lion business. We had a still closer brush with death in Wales, of all places. Bob was away from Zambia for family reasons, and we managed to hook up for a bit of Welsh birding. We were walking along a river, and came up to a single-track railway bridge. I think this was the Barmouth Bridge that crosses the river Mawddach, though this may be wrong. But never mind the details. "Much better birds on the far side," Bob said airily. "Never any trains here. Sunday, you see. Wales."

The bridge is about 900 yards across. So we set off. At about the halfway point a train appeared. It took up all the track, every inch, there really was no room on the bridge for anything at all except train. So we managed to get between the track-bed and the metal structure of the bridge that supports it, a pair of things like giant coat-hangers. The edge of this girder-thing we selected was L-shaped, forming a ledge about an inch wide. On this we rested out heels. We pressed our backs against the metal, using the pressure of our finger-tips on the track-bed to do so. And the train, two or three coaches long, rattled past an inch from our noses and passed on. Never paused for an instant, never hooted. We scrambled back onto the still-quivering bridge with our legs vibrating like egg-whisks from the weird pummelling we had just experienced. We exchanged a look. Experimented with a few monosyllables. I was far too shaken to give Bob the bollocking he richly deserved, or myself the reproaches I equally deserved; I've been with Bob enough times, I really should know better. And then – we walked to the other side. Not back. We never even discussed the matter. It only occurred to me later that we'd have to cross the bloody bridge again to get back. The moment we stepped onto solid land a merlin swept down almost at our feet and pursued something at drastic belly-scraping speed before vanishing: the best merlin moment I've ever had. "I told you it was better," Bob said.

It wasn't even Sunday. I realised later it was Monday.

Great merlin.

28

That Near-death Thing

"I wish," said the climber, "I was out there now. Out there with my fingers in a hold no wider than this pint glass and my feet smearing on nothing."

"What a dreadful thought," I said, with obvious sincerity.

"It's a wonderful thought," he corrected in soft Yorkshire. Not because climbing took you close to death. Quite the opposite, though he was a down-to-earth Yorkshire cragsman and he didn't get all mystical on me. I knew what he meant, though. I knew precisely what he meant.

A sports parachutist told me about the elation he felt when completing a jump. "It's as if you never realised before… well, how green the bloody grass is." He tailed off lamely. And clearly feeling that wasn't enough, he quoted *Jonathan Livingston Seagull*, of all things, one of the set texts of hippydom: "Perfect speed is being there." Perfect bollocks is selling a million, I said to myself with both literary snobbery and writerly envy, but again, I knew what he meant.

Not great at heights, me. Can't imagine myself up there with untold gallons of air underneath my feet, either holding on or jumping off. Always been fascinated by risk sports, though. Rick Broadbent, a colleague when I was at *The Times,* wrote a book about the Isle of Man TT races and called it, quoting one of the riders, *That Near-Death Thing*. Now, I'm no good at engines; can't drive a car, can't imagine racing motorbikes.

But I'll tell you one thing you always find with people who do risk sports: they will all deny, instantly and intensely, that they find the idea of death attractive, or even interesting. They will stress, above all things, their love of life: the way very intense experiences make life more vivid, more tightly packed with meaning. People who take part in risk sports say that they love life more than other people: they love it because it's wilder. For them, there's something ever-so-slightly sacred about the activities they take part in. It's usually deeply hidden, and even if found, well-disguised. But it's there all right.

I know at least something of this from personal experience. I'm a horseman, I've done many wild and crazy things on horseback; I've ridden like a madman in cross-country events and count these among the greatest and most important experiences of my life. I've had my share of falls and injuries, naturally. I'm saner and safer than I used to be, but I still love to be on horseback. I was, I stress, no braver doing this than I have been when walking with lions. I was there for the love of it. For the life of it.

In *The Second Jungle Book*, we hear how Mowgli loves "to pull the whiskers of death". So he contrives a wild, deranged and brilliantly successful plot to defeat the pack of depraved red dogs that invade the country of the wolves. And if you walk in the Valley you will get just a little taste of that. And it really is the stuff of dreams.

I think the striving for these experiences of danger and the urge to be in a place like the Valley, come from the same place. We have a need to be a little wilder. We sicken of too much comfort and safety. We seek the luxury of feeling unsafe. The savannas of Africa are a dance of life and death – and it's love of the former that brought me there and brings me back again and again.

Bucket List Ideas

Here are some examples of the sort of thing people put on their bucket lists: the things they would like to do before they die.

    See the Northern Lights
    Go sky-diving
    See penguins
    Swim with dolphins
    Plant a tree
    Go whale-watching
    Sleep under the stars
    Ride in a hot-air balloon
    Go scuba-diving
    Shower under a waterfall
    Go white-water rafting

And of course:
    Go on safari

So many of the things we have always longed to do are about regaining connection with the wild world and living a wilder life than we do now.

30

Lost Innocence

Almost at once my two months at Mchenja acquired all the characteristics of eternity, or at least of a child's school holiday that carries on and on and on, one unbelievable day of treats following another. I grew accustomed to living in the Valley: living inside the park. And every day, with the help of Bob and Manny and Perry, I learned more and I understood more. I wasn't the same person who had stared open-mouthed at the Valley just three years earlier. I had half-decent bird identification skills. I knew stuff about animal behaviour and ecology. I could look at the sandy floor of the river bank and notice that a lion had passed that way the previous day. I would note the central pad with three lobes at the back and four oval pads around it like little moons. And of course, no claw marks. In soft sand, which spreads wide at the soft, heavy impact, a lion's paw mark looks huge: that of a big male seems to have been left by a leonine monster the size of a carthorse, as Aslan sometimes appears in Narnia.

I was hardly a fundi. That's the Nyanja term for a wise person, a person who knows, a person of authority. When there was debate about where to walk, defer always to Perry: he was not only the scout and in charge, he was also the fundi. But then Bob was a fundi too, a man who knew everything about birds, and so for that matter was Manny, for all that he had only recently qualified.

But when the wind was southerly, I now knew a hawk from a handsaw, and what's more I could do it on call. I had begun to learn about the individual species and the way the bush functions as a living thing. I knew why the hyenas behave differently in the Nsefu sector; I knew something of how elephant society worked; I was fairly well up on the social life of lions.

I was no wide-eyed innocent, then. The first frenzies of pure craziness had long gone. Now, I have a theory about the big experiences of life: about love and about loss. Both begin in a kind of madness. Love is preceded by the state of being In Love. Being In Love is not a permanent state, nor is it supposed to be. I have – I hope we all have – known the insane frenzies of the early stages of love reciprocated. In Douglas Adams's *So Long and Thanks for All the Fish*, the couple step naked from their window and make love in the skies above London; well, we've all done that, in our different ways.

But then it changes. The great heresy of the 21st century is the belief that being In Love should be a permanent state and if it isn't, you've been cheated. You are no longer In Love because things have deteriorated and – worse still – you have a right to try and find that madness again with another. No, In Love is the essential precursor to a deeper and more meaningful thing which is, of course, Love: ever changing, ever developing, ever deeper. Being In Love prepares the way for something greater. We fall, we fall, but it's when we land that we love.

It's the same thing with loss. Grief is a transitional state, like the state of being In Love. And grief is followed by the deeper and richer state called sadness. Grief leads to a sadness you never lose or want to lose. You don't want to lose it because you never want to lose the person you loved and mourned. You don't feel less sad because you have moved beyond grief, but the madness of grief passes, at least for most people.

And more or less from my first day of my extended stay at Mchenja I was aware that my love of the Valley was quantifiably different to my love on that first trip, when elephants ate my hut and I drank with lions. Back then I was In Love all right, but that state was passing. I could now tell you what birds were calling while the lions and the elephants ate their fill, and have a good idea about why they were doing what they were doing. The love I had for the Valley was much less crazy now, and as a result, it was a much deeper and much more dangerous form of madness.

Walls of the Combe

A sacred place needs boundaries, to delineate it from the secular life that surrounds it. The combe I explored with Ralph was so deeply enclosed it felt as if we were the first people ever to step inside it. At Edenever the river was walled and roofed with trees. These places were cut off from the world: infused with a sense of privacy. But the privacy worked in our favour, not against us: not Private Keep Out, more like Private Stay In. It's everybody else that must Keep Out.

In *The Magician's Nephew* – what would now be called a prequel to *The Lion, the Witch and the Wardrobe* – the garden that contains the magical apple tree is surrounded by a high wall of green turf, pierced at one point by golden gates. "You never saw a place which was so obviously private. You could see at a glance that it belonged to someone else. Only a fool would dream of going in unless he had been sent there on very special business." Written on the golden gates in silver letters are the words:

> *Come in by the gold gates or not at all,*
> *Take of my fruit for others or forbear.*
> *For those who steal or those who climb my wall*
> *Shall find their heart's desire and find despair.*

The Luangwa Valley has its own walls. They run approximately north to south on opposite sides of the great river. To the east lie the Nchendeni Hills, rolling and rounded and of legendary steepness. You can find the name written on every map you care to look at, and it comes from a joke as ancient as the hills themselves. A beautiful young woman was walking up the hills with a heavy burden. She stopped for a rest, and after a while a young man came swaggering up the path behind her. She stopped him and asked if he would help her with her load. The young buck asked: "What's in it for me?" The woman replied: "Nchendeni!" It's a matter of cold and sober fact that these are nothing less than the Fuck Me Hills. I wrote a novel set in the Luangwa Valley but for the usual reasons I needed to change a few names here and there. Naturally I called it the Nchendeni Valley.

The wall to the west is much more wall-like. It is the Muchinga Escarpment, which rises from the valley floor with a due sense of drama. Muchinga means barrier, or boundary; I suppose at a stretch you could call it the Fuck Off Escarpment. (Manny agrees: "It says: fuck off if you are trying to conquer me.") The flat and gentle land of the valley floor changes, almost in a matter of strides, into a cruel vertical wall: a natural barrier.

One day Bob and I decided to have a bit of an adventure and climb the damn thing. In a Toyota Land Cruiser, rather than with pitons and ropes, but still not a straightforward business. The track that runs up the escarpment is hard to locate, and we got lost more than once and asked for directions more than once and got misdirected more than once. But eventually we found ourselves grinding slowly and with immense certainty up this astonishing track, the valley floor below growing more distant and less real with every leisurely turn of the wheels. Bob was an accomplished bush-driver, and he needed every bit of his skill and experience; we climbed upwards like a flame travelling the length of a very slow fuse. At one point we started going backwards a little – the slope was at that point about 45 degrees, or one in one if you prefer – but Bob corrected that somewhat noticeable fault. We crawled up to 2,000 feet above the valley floor, by which time the view behind us was hidden by trees and the impossibly abundant life below us. The herds of antelope, prides of lions, flocks of bee-eaters, pods of hippos seemed the product of our imaginations, our shared inventiveness. It seemed that I had reached a deeper understanding of the Valley by leaving it, in this rather special manner.

We returned late the next day, making a far more gentle descent into the Valley at a less precipitate part of the escarpment. It was memorable for the worst picnic of my life: at lunchtime Bob and I found that we had nothing to eat but a little stale bread and some chilli pickle. We took a slice each, duly smeared with pickle. We also had a beer each; these turned out to be a temperature at which it is more usual to drink coffee. Bob had even failed to park under a proper shade tree, so we consumed this gastronomic holocaust under the blazing October sun. Naturally by dusk, and still an hour from home, we needed further refreshment, so we dropped in at a bush-camp run by our old friends Craig and Janelle. Janelle, a devastatingly lovely Zambian lady, was educated in South Africa where she picked up the accent, with some high and lilting tonalities all of her own. She greeted us with the words: "You can have anything you like except beer." Except *byurr*. She was expecting clients the following day and had only half a dozen beers in stock. She generously agreed to swap our two remaining lightly-boiled beers for two cold ones, but after that we were forced to drink whisky. A couple of hours later we set off back to our own camp. About 30 minutes later, we returned to Craig and Janelle's place to explain that we couldn't find our way out. We kept making circles and missing the – admittedly not very obvious – main road. Craig climbed into his own Land Cruiser and we followed him to the road known as the 05 – Janelle calls it "the oh-fahve" – and then drove merrily home. It had been a good little adventure. I now knew where the boundaries of the Valley lay, and why they were so effective at keeping things out and keeping things in. Especially the latter.

32

Giraffe Crossing

Bob and I were parked up at the top of a Luangwa cliff when the geography of South Luangwa Park changed dramatically before our eyes. We saw a giraffe, a skyscraper bull, walking, in his stately superior way, across a wide expanse of beach – swinging his legs in that languid two-on-each-side gait that giraffes go in for. We expected him to stop, drink and then return the way he had come and we paused to look, because the complex mechanics required for giraffe-drinking always have a certain compulsion. We had clients and the chance to view the event and photograph it made them happy. The Valley does that sort of thing. It was indeed a fine sight, but the giraffe didn't stop; it didn't even pause. He walked straight into the river and carried on. He crossed the river with serene confidence, scarcely wetting his fetlocks. "Do you reckon we could drive that?" I asked. "Just what I was thinking," Bob said. And so suddenly the park on the opposite bank – the Nsefu Sector no less – was ours.

The following day, client-less, we gave it a bit of an explore. The only tricky bit was getting down from the summit of the banks onto the dry river-bed: in most places there's a drop of a good twenty feet and it tends to be rather vertical. But hippos make roads and they also make staircases: every night they leave the river and make a great circuit of their favourite route along their own homemade highways and then they come back to the river again to spend the day sloshing about and grunting. Elephants will often adapt these paths and stairways for their own journeys to and from water, and inevitably they extend and expand them. The natural and inevitable processes of erosion also lend a hand, and when the river gets high again it modifies and softens these gateways. So we found a place where we could get down and looped back to the exact place where we had seen the giraffe cross. Cautiously, Bob sloshed us gently across, and it was easy. There were no doors on the Land Cruiser – doors are for wimps – but the water didn't even get into the cab. We crossed dry-shod and drove along the beach until we found another staircase, one that Bob had identified from the far side. Up we went, straightforward enough, and then after the merest smidgen of bush-bashing we were back on the graded tracks and out there in the Nsefu Sector where the lush riverine bush gives way as the land gently rises to open plains where wildebeest lurk, with the great escarpment far beyond.

This, then, was the mighty Luangwa: a ditch you could trundle across as you please.

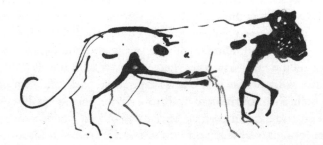

33

Jill and Jewel

Narnia is another paradise: a place where the animals talk, trees come to life and dance, fauns perform the snowball dance with the dwarfs. Jill talks to Jewel, the Unicorn – CS Lewis prefers the capital U – in the final book of the series *The Last Battle*. "'Oh, this *is* nice. Just walking along like this. I wish there could be more of *this* sort of adventure. It's a pity there's always so much happening in Narnia.'

"But the Unicorn explained to her that she was quite mistaken. He said that the Sons and Daughters of Adam and Eve were brought out of their own strange world into Narnia only at times when Narnia was stirred and upset, but she mustn't think it was always like that. In between their visits there were hundreds and thousands of years when peaceful King followed peaceful King till you could hardly remember their names or count their numbers, and there was really hardly anything to put into the History Books."

The Valley is a bit like that; the ecology of the savanna is a bit like that. It's not all about killing. It's not all about bloody encounters and ferocious mealtimes. Most morning walks are as easy and peaceful and glorious as Jill's walk with the Unicorn. You really don't see much: just beauty beyond all understanding. I have walked countless times and watched the pukus and the impalas stare solemnly back, and then sat on the banks of the river and watched the pied kingfishers hovering and the African skimmers skimming and looked on as the elephants cross the river as they pass their elephantine way from A to B, leaving high-water marks of black on the pale-grey of their great tree-trunk legs, and there's been no thought of violence and no wish for it: just hour after hour of unending loveliness and scarcely a thing to write in your notebook when you get back to camp, apart perhaps from, "This lovely valley really is really lovely."

That's true for lions as much as it is for antelopes. Lions love to sleep, and they love to sleep in a great big hot furry slobbering heap, often on their backs with all four paws sticking up in the air, so that they look not so much like the terror of the savanna as a bunch of oversized teddies. Hour after hour they spend like that, only moving to snuggle up a bit closer, and swat a vast affectionate paw on a neighbour, and they'll let you drive right in among them so that you could reach out and stroke them, and surely everyone who has ever had such an experience wants only to jump off the vehicle and roll and roll and roll among the lions and be an honoured member of the happy hot big-pawed community of aunties and sisters and nieces and nephews. They'll spend 18, 19 hours in a day like that if they have full bellies, and never trouble to look for a safe place; after all, a lion *is* a safe place. George Schaller did a great and masterful work about lions in the Serengeti, hour after hour of field observations, and many of them were about how long they all spent piled on top of each other in squirming self-indulgent slumber – so much so that he often found his note-taking professionalism fraying at the edges and a combination of boredom and empathy would send him to sleep as well. Lions are the most peaceful animals you could ever encounter.

34

Delicate Savagery

Impalas are perhaps the most beautiful mammals in the Valley or in all of Africa, and the most ignored: ignored for their glorious ubiquity. Impalas silence the cameras: cameras waiting for a large predator, waiting above all for a big cat. Sometimes it seems that to take your morning walk you must shoulder your way through crowds of impalas, as if you were going against the flow of arriving commuters to reach your train. That's in the dry season, of course, when no one can stray far from the river.

Impalas embody all the notions of beauty and fragility and delicacy and gentleness you could imagine from the word antelope: huge melting eyes, legs of pencil slimness, tiny hooves of liquorice. The males are horned but delicately – elegant headwear shaped like a lyre, more dandified decoration than weaponry. They run with a studied exquisiteness of movement, and will frequently throw in a vast extravagant bound. It looks like pure showing off because that's exactly what it is. It's an advertisement to pursuers: look how splendid and fit I am, I wouldn't bother chasing me if I were you. The subtext is more sinister: you'd be far better off chasing that smaller/weaker/older/younger impala over there.

On a night-drive the spotlight shines on the impalas as you pass, a bunch of them enduring the fearful darkness together: the eyes shine back so many and so close together that you seem to be looking at the lights of a distant town. Every mind behind each pair of shining lights knows that the night is the time of the leopard. The next morning you walk past crowds of impalas and they stare back with sad and wondering eyes like the angels in the hymn, keeping you just on the edge of their flight distance, so poised and so confident in their powers of flight that you could almost convince yourself that it was a moment of mutual trust and the human could lie down with the impala. Of course it's an illusion, it's all a balance based on the most ruthless computation of risk and distance and speed, a calculation as precise as if it has all been worked out for them by an actuary – and yet you can still savour the taste of paradise as you stare kindly at the impala and the impala stares understandingly back.

But wait. What's that hideous noise? It's wild roaring, deep, furious, echoing with all the ferocity that the wild world is capable of. Some hapless creature is being disembowelled just out of sight, or perhaps a group of carnivores are going through the most terrible round of in-fighting. Wrong. It's impalas. The males sparring: head-banging, lyre clashing against lyre and making the most hideous music. Impalas go in for dominance disputes of a spectacular kind. They crack heads with reckless aggression, they shove like rugby union prop forwards and they roar like bulls while they're doing it. Even when you're watching it you can't believe that such lovely creatures could make such an ungodly din. It's as if two ballerinas had gone cage-fighting and were making a howling success of it.

A male will seek to hold a territory and while he holds it he will be able to mate with most of the oestrus females that pass through. It's been described as a harem but that's a wish-fulfilment understanding from male observers. There's no real ownership. The females will move on when they choose, normally when they have grazed and browsed the place off. The male tries to keep them on his own patch for as long as possible, naturally, and he does so by herding them and by seeing off any rival invading males. It's a full-time job, one that doesn't allow much time for eating and drinking, and so inevitably the male weakens and will eventually lose a fight and his patch. So he will join a bachelor herd, a gathering of male impalas who aren't holding a territory – some too old, some too young, some too defeated – and attempt to eat himself back into strength, so he can take on another male, win another territory and get back to those luscious females. These delicate beasts are warriors: ferocious as any lion in all but diet.

35

Sensitive Types

Those macho, roaring, territorial impala rams don't have it all their own way. And you don't have to be a territory-holding ram to cover one of those sexy ewes and propagate your genes. There is scope for, if you like, the more sensitive and artistic sort of impala male to find a female and to propagate his artistic and sensitive genes. He does so by hanging around on the fringes of another male's territory, eyeing up the females at a discreet distance and being eyed back. When the alpha male is busy chasing out a rival at the far end of his territory, he can come in and have his way with a willing oestrus female. Copulation by sub-dominant males is something you find all over the wild world, and the zoologists have a technical expression for it.

Sneaky fuckers.

36

Manhattan White

I was at an open-air concert on the downs, perhaps a hundred yards from Clifton Suspension Bridge. The year was 1973. I had withdrawn from the crowd of friends I came with because even then I disliked loud noises. As I walked round the fringes of the gathering I was approached. You could tell by the manner of approach what was coming next, but the details turned out to be exceptional. "Do you want to score some grass?"

"Maybe. What you got?"

"Manhattan White. Really good stuff."

"Sounds it."

"No, really, man. It grows in the New York sewers. I mean it's really hot and fertile down there. And the heads in New York, like they're all really paranoid and every time someone knocks on the door they think it's the pigs come to bust them and they flush their stash down the bog. And it goes into the sewers and the grass there is always full of seeds and because it's so fanatically rich down there it grows like anything. Only because there's no light it grows pure white."

I have been haunted by this tale ever since. I love the idea of the fecund sewers beneath the world's most citified city, pullulating with life, vast ghostly white marijuana swamps in which giant white blind alligators frolic at their ease: a place of stinking dereliction transformed by the might of nature into a sacred combe of sorts far below the city. I was aware even then that photosynthesis doesn't take place without the help of the sun, a fact that explains why plants are so rare in sewers, but there was something more than dealer's bullshit here. No, here was a fantasy of secret beauty and wonder, a dream of a place within our reach, if only we knew, a sort of paradise that we built in spite of ourselves by flushing drugs and unwanted animals down the lav. From human fear and human callousness a little Eden springs up: I could almost see myself there with a hippy Eve, laying aside her beads and her Indian kurta to join me dressed as Eve should be; come, love, come, come with me, down to the eternal white marijuana swamps of Manhattan where the human shall lie down with the alligator and we shall all the pleasures prove.

I wish I'd bought some now. But instead I just smiled and walked on, parting from the dealer with a politeness of the time.

"Far out."

37

Secret Sacred Monsters

Perhaps the alligators of New York are the first modern urban myth. The tale dates back to the 1930s, at least, and tells of how New Yorkers came back from holidays in Florida with cute baby alligators for the kids to look after, and when they got too big – say, two feet – they flushed them down the can, or john, whereupon the alligators took to the sewers, lived on rats, flourished in the great damp warmth beneath the city and grew to enormous size. There have indeed been alligators in New York. An eight-footer was found in Harlem in 1935, but it was probably the inadvertent passenger on a ship. It was slain by teenage boys. In 1959, a book called *The World Beneath the City* by Robert Daley was published. In its pages, a former sewer worker, Teddy May, claims to have found a colony of alligators in the 1930s and exterminated it. The story has certainly grown with the years. The alligators are often said to be white, because of the lack of light. In some versions they have mutated drastically in response to chemicals in the waters of their urban caverns. The fact is that herpetologists – people who study amphibians and reptiles – agree that the sewers are too cold to sustain alligators for 12 months of the year, and anyway, they are too filthy and polluted. That hasn't stopped the legend of giant blind white alligators thriving beneath the city. It seems that part of us wants it to be true.

The same yearning explains the persistence of stories and sightings of hidden beasts, technically called cryptids: the Loch Ness Monster, Bigfoot, the yeti or abominable snowman. It is a fine and fertile area for the credulous, the hoaxers, the imaginative, the obsessive and the slightly mad. There is even a name for it: 'cryptozoology', a pseudoscience based on anecdote, fakery and willingness to believe... and even as I write these cold words I feel like a spoilsport.

Britain is plagued by secret cats. They are the seldom-seen ravagers of sheep, a mortal danger to rural children, sometimes sighted in the distance by early morning folk, sometimes, still less reliably, by late-nighters. These cats are almost always black. Fact: melanism in big cats is not at all common. Possible explanation for the blackness of Britain's cryptocats: a lot of domestic cats are black, and – broadly speaking – black is the only colour that big cats and domestic cats have in common. As every birder knows, it's extremely hard to judge size; we've all been fooled by distant eagles that turn out to be much-closer hawks. So here's George Monbiot in his book *Feral*: "If a breeding population of these animals existed, hard evidence would be abundant and commonplace. Its absence shows that there is no such population. With the possible exception of the very occasional fugitive (almost all of which have been quickly caught or killed and none of which is black), the beasts reported by so many sober, upright, reputable people are imaginary."

So why are we so eager to imagine them? Monbiot postulates an atavistic "mental template" that encourages us to see cats, like the mental template that makes us see faces in clouds and cracked ceilings. He suggests that this cat template is an ancient warning device from the time our ancestors first walked on the savannas of Africa. I like that, but I think there's something else involved as well. I think the prevalence of Britain's imaginary and hallucinated Manhattan-white-inspired big cats also reveals a nostalgia: a deep human need for large and dangerous beasts. There's a part of us that wishes we were back on the savanna, or at least, living a less circumscribed and concreted life. A nostalgia for peril is part of the modern human condition. In these times of peace and plenty, what we really long for is wildness. We yearn for our enchanted combe, but we prefer it to be a little dangerous.

38

Some Famous British Big Cats

Surrey Puma
Fen Tiger
Shooters Hill Cheetah
Beast of Exmoor
Sheppey Panther
Galloway Puma
Beast of Dartmoor
A panther was seen by 14 people in Shotts, North Lanarkshire, in 2011.
A lion was seen and photographed in Essex in 2012; Mrs Murphy recognised it as her (admittedly substantial) cat Teddy.
A big cat, black of course, was seen near Wrexham in 2013.
In 1983 the Marines were sent in to shoot the Beast of Exmoor, said to be responsible for the death of 200 farm animals. No shot was fired, though a number of sightings were claimed. The commanding officer noted the creature's considerable, near-human intelligence, saying that it "always moved with surrounding cover amongst hedges and woods."

39

Being Prey

There was me and Bob and Perry – Perry, as always, leaning on his rifle in that nonchalant, boneless way of his. Also a client or clients; I can't remember. We were on foot. We had walked across a brown, dry, crackling area of open plain because we had seen a few vultures perched at the top of a lone tree and we knew what that meant. We were right, too; beneath the tree the 12 core members of our lion pride were lying in a ragged rosette around a dead buffalo – killed not long before, judging from the ferocity of their appetites. So we stopped at a discreet distance, maybe 40 yards, to observe, to be a part of this primordial scene. We could hear the sounds of the banquet, the avid shearing work of the carnassial teeth.

I'm not sure what we did wrong. Maybe we stepped a little closer, maybe a client spoke over-loudly, maybe the too-bright garment of one of our guests caught a leonine eye. Either way, everything changed in an instant of time. One of the lionesses, the boss female, was on her feet and giving us a death-ray stare. I could feel the golden eyes like a physical force. Had we all run, there's little doubt that she would have chased, and the others would have followed, but let's not overdo the danger. Had we all run in a body *towards* the lions they would have fled. Almost certainly. Well, very probably. Lions, like everybody else out there bar a few humans, don't believe in false bravado. Living to hunt the following day is what counts, even for the most ferocious carnivore in the bush.

But all the same, it was a powerful moment. One of those vivid reminders that you are not, after all, a being with a unique status among our fellow-animals. You are perambulating protein just like everybody else, just as the horizontal buffalo once was: well, still protein all right but no longer perambulating.

Humans have been hunted and eaten ever since they started walking upright on the savanna: there's nothing unusual about it, still less anything depraved or diseased or unnatural. Quite the opposite. It just seems unnatural for us humans to consider ourselves as part of – rather than observers of – the natural world. We've always been prey: here was a blazing reminder. The boss-lady, she was angry, tense, ready to do what was right for herself and her relations sharing her meal. So far as she was concerned, what happened next was her call. So she stood still and we stood still and we all waited for the stand-off to unravel. Eventually, she sat: stiff, upright, like a cat on a mat, front paws close together, almost touching, while the rest of the pride, their heads lifted from their meal, added their stares. Once again, we were not observers but participants. Once again we were not proud puffed-up members of the race of Leonardo and Darwin and Joyce and Bach: we were part of the continuum with buffaloes and antelopes. And that's why we walk. That's why we walk in the bush when we could take much better photographs by riding in a vehicle. We aren't spying or watching or observing: we are participating.

We waited, at Perry's mild but persuasive suggestion, until the boss-lady was back down on her breast-bone and eating once again, even though her eyes constantly returned to us. We retreated at a courteous 45 degrees: not bolting in any obvious fashion, clearly not advancing. And slowly the tension drained out of the bodies of humans and lions and the world was once again full of peace. And among the clients, the time of compulsive talking had begun. I remember John Coppinger, who runs a camp further upstream, telling a story of some scrape he and a friend had got into, and after he had told us about their escape he added the memorable line: "Man, we were giggling like *schoolgirls*."

40

Nsefu

The river was still narrower. You or I could almost leap it. Bob Beamon could have crossed it dry-shod. It had given up flowing. The Luangwa was a lukewarm ditch. So for a week or so, we were able to wade across every morning, no need for giraffes to show us the way. I rolled my trousers to the knee and, boots and socks in one hand and binoculars in the other, walked the brief distance without anxiety. Obviously we looked for crocs, but there wasn't enough water to hide a two-foot alligator bussed in from Florida. And so we would walk through the subtly different landscape of the Nsefu Sector.

It's a place that always seems special: a special place within a special place. Like Digory's walled garden in Narnia, it was a paradise within a greater paradise. Partly that's just because it's the other side of the river and for most of the year beyond easy reach. And partly it's because Nsefu means eland and elands have always seemed to me the embodiment of peace and tranquillity. Diagnostic signs: they are very big, very pale and very distant. The bulls weigh as much as a tonne, and stand five feet high at the shoulder; a standing eland can look down on a standing man. They look a bit like antelopes trying to be cows, but it seems to me that where cows have vacuity, elands have serenity.

Elands are never close. They aren't fast, like pukus and impalas and zebras, so they have to be clever. And hyper-aware. Impalas will let you come within a cricket-pitch or so before they move away, trusting in their speed, their knowledge of the country and their safety in numbers. Elands have no such luxury and so prefer a distance of two or three hundred yards. With that established they will stand solemnly and observe. I can see them now, backlit, thin scrub, half a dozen of them staring back, and even as you stop and raise your binoculars they are drifting back: one-tonners drifting back like pale wisps of smoke, not retreating but dissipating: dissolving into bush.

I spent half a day in a tree hide, alone, staking out a waterhole. It was a long, slow, quiet time and the sun crawled grudgingly across the sky. It was getting towards dusk when I heard a sort of Geiger-counter clicking away behind my tree, just out of sight. So naturally I knew it was an eland bull long before I saw it. This click comes from the tendons in the legs and some say the depth of the sound is indicative of the eland's size and therefore a dominance signal, though I have heard others explain it as a here-I-am sound that keeps herds together in poor light; elands are lovers of dawn and dusk. And here at dusk the lone eland bull came to drink his fill and rest before clicking his way back into the bush. And I, observing unobserved, was back in Eden again. It seems to me that no depiction of Eden is complete without an eland or two: as much a part of the place as the naked human couple and far more convincing than vegetarian lions and tigers. In their generous size, their distance, their wisdom and their slowness they seem to embody content: as if danger, as if risk, were a million miles away and things nobody could ever desire – still less seek out.

41

Morning

"Good morning, Simon. Your water is ready."
The softest and gentlest voice in Africa woke me.
"Thanks Denis."

So it was half-past five. No, really, I wasn't going to get up yet. Not until I had identified, say, ten birds on call. African pied wagtail, ground hornbill, Cape turtle dove, African fish eagle, hadada ibis, greenshank, Egyptian goose, barred owlet time he was in bed, Heuglin's robin, white-crowned plover, damn.

Wash, clean teeth, take turn at long-drop, sorry if that's too much detail. Dress: jocks and socks, cotton trousers, ditto shirt, Timberlands, hero hat. Binoculars in hand. "Morning Jess, morning Bob."

"Wood sand." I looked out. "Behind the spoonbills."

"Got him."

Wheeze-honk of hippo in the river before us. Shrill whistle of puku, coming down to drink. I continued scanning. "Is that a purple heron? About 300 yards off?"

"Ah yes, well done. Though actually not. It's –"

"A Golly-at?"

"Precisely." Meaning a goliath heron: the biblical name is pronounced that way in Zambia. I once met a scout, a man of impressive size, who was named Goliath and of course pronounced Golly-at.

From the trees behind camp I could hear an orange-breasted bush-shrike singing a variation of the first bar or two of Beethoven's Fifth. The baboons squabbled suddenly, loudly, briefly. They're not quarrelsome, they're addicted to making up. They reinforce their way of life by making peace all the time. It works very well, but they need to have regular fallings-out so they can make up again. It's a system that's been known to work with other primates. Not us lot at Mchenja, though. We were well content with each other. The current clients were nice and loving it all; they emerged from slumbers and ablutions and gave greetings and sought coffee. A silence, companionable enough as we all scanned the river, in search of birds, in search of something else, too. Either way, the prospect before us gave us what we sought.

I've never been much good with hand-tools, but even I have known an occasional moment when you get a saw going nicely and it makes a series of resonant in-out rasps. That's the sound we all heard quite clearly from about a hundred yards off. "Leopard!" And clients entranced and joy all round.

Manny, who always stole an extra few minutes in bed, arrived with one of his easy smiles. "Must have been the one we saw last night."

Perry was now here with his rifle, a mild call to order. "Those lions will probably still be there," he said.

Time to go then. Perry in front, Manny second, me at the back. Bob was having a morning off; Jess went back to work at the camp. We were taking another morning stroll across Eden: to see if the lions were lying down with the lamb under the old apple tree.

42

A Good Old Lovey-up

The tiger and the leopard are engaged in a merry old transpecific rough'n'tumble, while two species of macaw look down from the branches. There's a bunny and a puppy-dog on the ground. There's an American wild turkey, of the kind I have often drunk on the rocks, and a golden pheasant. I can also see a goldeneye on the brook and a surf scoter before it. There's a mute swan, a heron, a white hart that looks as if it has just stepped down from a pub sign, and a camel. Lion, cow and ostrich gaze benignly at the naked couple beneath the bird-thronged tree, while a peacock seems to be auditioning to be a duvet. There's a cat having a good old lovey-up around the calves of the standing woman. All this from one half of one of the greatest double-acts in the history of painting, Jan Breughel the Elder.

His colleague, Peter Paul Rubens, painted the rest. He gave us a horse looking over the seated man, and the snake, a python of some kind, spiralled comfortably around a branch above the couple's heads. And of course, he also painted Adam and Eve: Adam gazing, as if transfixed, by the admittedly excellent breasts of his Eve. Eve herself, apple in hand, is rocked onto her heel with her shoulders thrown back, so that she seems to be offering her crotch to the avid man before her. As well as an apple.

It's the moment before. It's like the *look-behindjer* moment in a pantomime: the greatest *look-behindjer* moment in the history of humankind. Don't eat it! My father's father – my grandfather, I know, but he died long before I was born – used to tell of an occasion when he was watching *Hamlet*; as Gertrude reached for the poisoned cup, someone in the audience gave an anguished shout: "Don't drink it!" You feel like shouting similar advice at this painting of the last two people on Earth, who had everything they could ever want or need. One second later, this scene of bliss will be destroyed forever. Those big cats will fall out with each other and all the other creatures will fear them and flee. The birds will disperse, thunder will strike, and the naked couple will disobey God's one command to them. Be happy, he told them, but they refused, knowing better.

This, then, is the last recorded moment of paradise: the last second of complete happiness that humankind ever lived. We, who observe, know that misery is just a second away – and yet we can't help but be seduced by the joy of the suspended present. Here there is beauty all around in every animal form including that of humans, eternal freedom to indulge in what Blake called lovely copulation, bliss on bliss, and a free and easy kinship with everything else that lived on Earth – for it's quite clear that everything that lives is holy and that life delights in life. That's more Blake, of course.

The Eden myth is of course a terrible warning of what we humans have lost: of how humanity had it all and lost it all. It can be seen as a warning of where nakedness and sexuality will get you, or of the way that man was, and forever is, brought down by woman. But the hell with that. That's not there, not in this painting. As you look at the painting today – and I bet it was much the same as you looked at the painting when it was finished in 1615 – what strikes the eyes and the mind behind them is not loss but joy: a lovely naked person in a lovely naked place surrounded by lots and lots and lots of lovely animals, and what's more, these are not joys gone forever but joys we can still understand, and understand because they are still – at least to an extent – available. In this picture of humanity and all-too-human fallibility, do we not also feel the deep joys of the non-human world: a feeling that joy is deeper in this most marvellous place because there are other animals there besides us humans – because humanity is not and has never been enough for us? There is a term coined for this: biophilia, the human need for contact with non-human life. It was not needed before the 20th century, when there was contact enough for such things to be taken for granted. But now biophilia is something we have half lost: to be revived when we smell a rose or pat a dog or see a passing skein of geese from the window of the train that takes us to work.

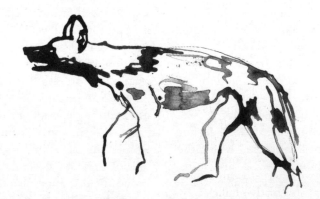

43

Lovely Leah

The point about a sacred combe is that it lies outside the common run of life. It's not where you live, or even where you want to live. It's part fantasy, part unattainable goal, part escape. It wouldn't be sacred if you were there all the time. It can't exist in your imagination without a sense of otherness.

Our clients landed at Mfuwe airport. The idea – and we really did manage it quite often – was to get to the airport before the plane landed, stand there with our sign, 'Savanna Trails', and an expectant expression, and then meet and greet and take them back through the park to Mchenja, the ebony glade, the river and mile upon mile of wonders. Whenever we could, we dropped into the Croc Farm first. At one stage this had indeed been a farm for crocodiles, but the market for crocodile skin failed and the place was – back then, before it opened as a tourist lodge – pretty quiet. It had a bar though, and there we would drink a couple of cold Mosis, colder than we could get them in our own fridge, which was run on paraffin. But for me the real reason for the visit was Leah. Leah was kind and delightful, and sometimes when she saw me her face would light up with joy, and that stirred my heart most strangely. "I have something for you!"

"Oh Leah, how wonderful you are!"

The Croc Farm owned the only fax machine in the Valley: an incomparable luxury, a technical advance that was almost in the realms of science fiction. And on those joyous days Leah would give me half a dozen or more sheets of fax paper covered in my wife Cindy's beautiful Italic hand, and I would rejoice in news from home with love and kisses at the end of it. I would give Leah a sheaf of A4 covered in my own rather more speculative Italic scrawl, and a small financial transaction would take place. And I would be overjoyed, knowing that my life was still working.

Had I been 20-something and uncommitted, my history in the Valley would have been very different. As it was, I went to Asia instead. I spent four years there, during which I met a beautiful girl called Cindy. But now, much later on, I was profoundly married and wanted no life without Cindy. I want, I need, to make it quite clear at this point that my love for the Valley was in the context of a rich life in England that I had no wish to change. It's just that this life was still richer for the existence of my sacred combe, for the possibility – thanks to the extraordinary and unending generosity and understanding of my wife – of spending some time there, and on this occasion, spending eight weeks of my life doing something it seems I had always wanted to do – had wanted to do long before I knew about the Luangwa Valley, in fact.

I don't spend all my time in England wishing I was somewhere else. Far from it. But the sacred combe enriches all my days, no matter where I happen to be. It's a contradiction. Everybody's life is full of them.

44

Doing the Knowledge

Manny was born not far from the Valley. The Valley was his university, where he worked for some years with Bill Astle, who was mapping the Valley's vegetation and geology; an extraordinary piece of work, still relevant today. After this Manny went into tourism and became a guide. To make any comment at all on the excellence of Manny's spoken English would be absurdly patronising, and his knowledge of the bush and its wildlife is profound. All the same, taking the viva exam for your walking license is a pretty nerve-racking business. Back then you were examined by a board of old-school all-white old Valley hands. They were, though, delighted to have such an exceptional person as Manny, for this was obviously the future of the Valley. And as they tested Manny and his answers revealed greater and greater depth of knowledge, one of them asked: "How do you know so much?"

I don't know what answer he was expecting, but he deserved all he got. Manny smiled sweetly: "I have read a lot of books," he explained kindly.

So have I. Also, like Manny, I have spent time in the bush with a lot of very knowledgeable people, Bob and Manny prominent among them. Adam and Eve were not supposed to eat from the Tree of Knowledge, but knowledge spoils nothing in the Valley. The only thing that knowledge can destroy is ignorance. The Valley has mystery all right, but knowledge doesn't destroy it. Rather, knowledge revealed more and more interesting and more profound mysteries, and these mysteries I talked about for hour after hour in the vehicle with Bob and with Manny, and with Jess back at Mchenja. The Valley consumed us all.

45

How We Destroyed Eden

The idea of the Garden of Eden is as old as recorded history. You can find Eden, or somewhere similar, in Greek, Sumerian, Mesopotamian and Persian mythology. You can also find Eden in the Quran: Iblis refuses to bow down to Adam and Eve in the Garden and so Allah turns him into Satan. In his new guise Satan tempts Adam and Eve to eat the forbidden fruit that will give them immortality.

It was the human invention of agriculture that allowed us to begin our conquest of the earth. Agriculture freed us from the uncertainties of the hunter-gatherer life and allowed us to take control of our destiny. The price was that we became tied to a place and committed to the endless unforgiving rhythm of hard work. We found freedom, and the price was slavery. When life is good for a hunter-gatherer, there is no need to work for more than two or three hours in a day at gathering; while hunting is a kind of glorious sport with a wonderful protein bonanza when you get a good result and the prize of hero status for the champions. In contrast, looking after the land is back-breaking labour. We traded fun and leisure and uncertainty for greater certainties, a longer and more secure life, a better chance of becoming an ancestor, and it's half-killed every one of us ever since.

And I suspect that as we did so, we retained – and retain to this day – nostalgia for a time, half real and wholly imagined, when life was easy and perfect and generous, when nature and humans were all on the same team. Sex all day, food dropping from the trees, nothing to come and eat you, the weather as kind as the lions: oh brave old world that had such creatures in it. It wasn't Eve or the serpent or the apple that destroyed Eden: it was the digging-stick.

Now we live in the 21ˢᵗ century and most of us have little experience of agriculture, still less of hunter-gathering, and yet we all still work our bollocks off, males and females alike, and still we have the same strange nostalgia for the beautiful, the perfect, the easeful, the generous place: a thing we seek in "luxury" holidays and in dreams of retirement. Our conquest of the planet is complete: almost everything that can be tamed has been tamed. It seems that the last luxury – the only true luxury we have left – is wilderness.

46

The Sin of Pride

"They are crazy, these lions," Perry said, and not without pride. They had killed again, another buffalo. The effete lions of the Serengeti and the Masai Mara kill wildebeest and zebra; the Luangwa lions specialise in buffalo: the most dangerous prey of all. They can do so because they are fast and they are powerful and they are many. You can't go after buffalo without picking up injuries, and that would be disaster for a solo hunter. A leopard would never take an avoidable risk because the pardine strategy is to be perfect. Any falling-off is potentially fatal, because if an adult leopard can't kill for himself – or herself – then starvation must follow.

But not our crazy team-handed lions, all lying together in a great big furry heap with their bellies full of buffalo. Often you saw one with a fresh scar, another with a new limp. Such an injury means that your hunting abilities are compromised, but if you can keep up with the pride you can still eat and if you can still eat, you can recover. And there's no resentment about sharing a buffalo with a stricken sister, partly because they're all related and the ties between them are very close, and partly because there's so much damn meat on an adult buff that a passenger or two makes no difference.

There were 12 of them in the core of the pride. Lions are an archetypal male symbol – power, strength, ferocity – but a pride of lions is anything but. It's a deeply female thing. For all that fearsome weaponry, lion society is all very hippy and bosomy, this one in particular because the alpha male was a near permanent absentee. There were three young males in the pride, with their manes starting to come, so they looked as if they needed a shave. The alpha male would have them out on their furry ear the minute he saw them taking more than filial or fraternal interest in any of the females. The pride of 12 was a girl thing, and the female animals in that pride were sleek and fit and hard as hell. So they should be: this was their time of year. The Luangwa River was dwindling faster every day, but all the other drinking-places had dwindled much faster. For every large mammal in the Valley the river was life; and before the rains came again, for a good few of these it would also be death. Every large mammal in the Valley needed to be within a few miles of the river, and that's why you could see such prodigious quantities of animals every single day. That explains why the carnivores were in clover: they scarcely had to rise from their slumbers to hunt, for the game came to them. It had to. All a lion needed to do was to stay within easy commuting distance of the river and wait for another meal to walk up. It's a time of year that makes for the finest game-viewing in Africa, and if you stay there longer than a week full of wonder you begin to experience a slow and terrible revelation of desperation and cruelty.

Not for lions, though. They were arranged round the latest fallen buffalo, eating their fill once again before ambling off to the shade. Lions have a special way of lying down at such times: they act like puppets who have had all their strings cut by a pair of gigantic invisible shears. Flop. Instant peace. And the lion shall lie down with the other lion and they will companionably lick the blood from each other's faces.

We saw this same pride of lions almost every day. They didn't need to move about much, save when they felt like it, and lions are inconsequent and capricious creatures who often seem to do stuff just because they happen to fancy it. Lions in a sense were our business: give your clients a good time with a few lions and they go home happy. It's only clients short of lions that get fed up with the long-drop or the rather curious arrangement we had for showering. And for most of them, the lions were kind, generous, obliging and static. And hungry. Always that.

47

Ring a Dong Dillo

When it comes to *Lord of the Rings* I line up with Hugo Dyson. Dyson was, along with CS Lewis and JRR Tolkien, a member of the Oxford literary group called the Inklings. He played an important role in Lewis's conversion to Christianity, which means that he's partly responsible for the Narnia books. So there were the Inklings all gathered together in *The Eagle and Child* with glasses full of ale before them and their pipes all bubbling away nicely and there was Tolkien reading the latest chunk of his great work to the company when Dyson interrupted with a devastating critique of the entire work. "Oh fuck, not another elf."

I read the book, of course. It was a compulsory text in the late 60s. But very little of it stuck. The one bit that stayed behind to haunt my imagination, more or less against my will, was the episode of Tom Bombadil. Tom is an enigmatic figure, benign but not without a sinister side. He has a rather tenuous relationship with the actual plot, making him unusual in such a densely plotted book. They left him out of the film: too confusing.

Tom is a sacred combe man. He lives in his own little bit of Middle Earth, an enchanted acre where no harm dares to come a-calling. He's a sort of living spirit of rocks and stones and trees: "master of wood, water and hill". He speaks in rhymes and verbal odds and ends: "ring a dong dillo" and so forth. He lives there in apparent bliss, in a delicious place with his (apparently) much younger wife, the delicious Goldberry, who's a bit of a doormat. When the hobbits visit him, Tom tries on the ring of power – and glory be, it has no effect on him whatsoever. And that's against the deepest premise of the book. He seems immune to the ring's powers of temptation, immune to its deep wickedness, immune to the promise of exponentially greater power.

This is all so weird that in the later council scene, the wise people of Middle Earth give serious thought to the idea of handing the ring over to Tom as a way of disposing of the problem of evil once and for all. But they don't take that course. They believe that despite Tom's immunity to the ring, neither Tom nor his secret, sacred place could hold out indefinitely. Not once it was entirely surrounded by the forces of destruction. A sacred combe is also defined by its vulnerability.

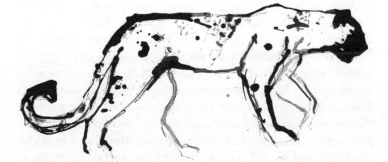

48

The Spotlight Kid

Rodgers Shawa was a trainee guide when I was at Mchenja. It was his duty and pleasure to wield the spotlight on night-drives, and he was pretty good at it. Spotlighting is the best way to see leopards, and Rodgers found them about one night in three, which meant that most of our clients left the camp filled with visions of glory and beauty. But when Rodgers was on leave I would take the light myself and spot for Bob, just like a real guide.

Sundowners. We stopped as always on a cliff overlooking the river, and we all took drinks. A Mosi, for me and a handful of fresh peanuts, the sun hastening down the sky as if it was late for an appointment on the other side of the world. And the day shift clocking off and the night shift clocking on: a scops owl, the wail of the water dikkop, called the flat battery bird because the sound slurs and flattens and deepens as if its motor were running down. In no time, it was twilight: a hyena whooping from across the river to add to the drama of it all. The first Mozambique nightjar's cog-slipping whir. Bob fixed the spotlight's leads to the battery, and I would take a stand directly behind Bob, feet on the spare wheel for added height, left hand grasping the rollbar, right hand making experimental casts with the pale beam; by the time the clients were all on the vehicle it was glaring out like Eddystone Lighthouse. Let's go.

Oh, what great sport this is. It's the primal emotion of nabbing something; better still, picking it out from deep cover and revealing its secrets for a fleeting moment. You've got to be careful and considerate, and a guide that fails to do so will have his licence suspended; I know people that's happened to. Elephants and hippos hate the spotlight, so never catch them in a direct beam. You can dazzle an antelope and make him unaware of danger and set him up for a leopard, and that's not on. You can also interrupt a leopard hunt and spoil it for him, and that's probably a worse crime, since most leopard hunts end in disappointment, at least for the leopard.

But do it righteously and you cause little disruption and give a great many people a great deal of pleasure, and make them a little more inclined to visit and to protect these wild places. Not every park in Africa permits spotlighting – most don't – so it's another little edge that the Valley possesses. Here you get to know and understand the night nearly as well as you understand the day, making it a much deeper, much richer experience than you can find elsewhere in Africa.

But I was talking about me. Hosing down the trees and sweeping the open savanna with the spot. What you're mostly looking for is eye-shine: the light reflected back from the tapetum, the reflective surface at the back of the eyeball that most mammals – though not us primates – possess. If there are a lot of lights together, pass on briskly: generally nervous night-time gatherings of impalas or puku, with lights like a small city. You're looking for eyes that come in pairs. Genet: a favourite of mine, spotty with an absurdly long ringed tail and a pink nose like a glacé cherry. Civets, famous for those vast piles of turds. White-tailed mongoose, a nonchalant midnight rambler. Honey badger, with a great reputation for ferocity and fearlessness. I'm not sure I believe the stories of their propensity to go unerringly for your balls when you corner them, but I don't plan to put it to the test. They're impressive beasts.

I remember a night when there were just two clients on board, a young Italian couple. You always want to find something special for the nice people: in a way their niceness adds to the pressure on you to find something wonderful. The more they love the bush, the more you want to show them its wonders. If they are surly and indifferent and complaining, you don't really mind if you can't find them so much as a genet, even though a sensational leopard sighting might shut them up for an hour or so. That's the bush, Sunshine, wild animals are wild, and if you don't like it, you can – but that's enough of that.

I found a pair of eyes, no, two pairs of eyes, no, hang on a minute – let's go in, Bob, there, under the overhang of the bank. And what a sight I rewarded these nice people with: three lionesses and four cubs between them, the cubs engaged in a full-on romp. They pounced on tails; they ambushed each other; they rolled about in mortal combat; they flopped down for a sudden rest and then got ambushed again in their turn and returned to the fray. At one point, one of the adults rolled onto her back, picked up a resting cub and tossed him into the air: the adult initiating the play.

These were part of the same pride of 12, an outlying group of related females, two of them lactating – lion cubs will boldly suckle from any lactating female in the pride, and the females will tolerate them, in the certainty that they're all relations stuffed full of genes in common. The third was perhaps a grandmother, perhaps a sister, anyway she preferred to hang out with all this young life rather than socialise with the rest of the pride. The social life of lions is based on a single principle: one of these days we really must get organised. Let's make a start early tomorrow. Well, mid-morning, anyway…

And so, having been vastly impressed by my own brilliant spotting – a blind man would certainly have missed them – we turned back towards camp, with me still sweeping the spot. Just one more: a genet high in the tree's fork, looked down with apparent approval as we paused for a moment to savour his beauty. Surely, I thought, there was room in my suitcase for him when the day for my return finally dawned.

49

Wild Centaurettes

The sacred combe is no idle fancy of mine. You come across it all the time, even if all you can find is a mourning for its loss. It really is something common to us all: that dream of a special place that extorts from us a kind of reverence. It is, for example, the reason for the game of golf. Golf I dislove; I speak here as one who has pursued the trade of sportswriter for getting on for 40 years, bar the odd break. I can seldom resist the opportunity to tease golfing people – not a real *sport*, is it, sport? Not a lot of physical risk involved, is there? I mean, you don't even have to run about etc etc. But perhaps the real reason I don't get golf is that I already have a sacred combe of my own.

Golf takes place in a green place, normally within easy commuting distance of a big city: an accessible Eden. It is traditionally a place where men get together with men: a prelapsarian vision of the world before Eve spoiled everything. Some golf courses even have productive areas for wildlife – Aldeburgh Golf Club in Suffolk had breeding woodlarks before their neighbours at North Warren nature reserve; I put that in for fairness and balance. But the prevailing vibe on a golf course is cosiness: Nature Lite.

I have a weird fascination for the annual event called the Masters Tournament. It takes place in Georgia in the United States and it's regarded as one of "the majors", as if it were a guest at Fawlty Towers. Its reputation as a truly special place is complete. One of the most cynical men I know covered the Masters and came away with every item of merchandising he could buy. He had the windcheater, the hat, logo-encrusted. He, of all people, bought into the myth of Augusta as sacred combe. As for real golfing people, they go weak at the knees over the Augusta National Golf Club, where blacks were finally admitted as members in 1990, a former rule stating that all caddies must be black was rescinded, and finally, with breathtaking audacity, they permitted women to join as far back as, er, 2012. (Fancy wanting to be one of those women ... Still, I suppose someone's got to do it. And anyway, it's still way ahead of a good few eminent golf clubs in Britain.)

Augusta is supposed to be supremely beautiful: a dream of Arcadia in which the slow, desperate duels of golf take place beneath the eternal blossoms of the world's loveliest trees. It's ghastly. The green of the greens and fairways has an eerie iridescence, lacking the broken colours of the natural world. Every hole is named for a flowering tree or shrub: Magnolia, Juniper, Camellia, Pink Dogwood, Flowering Peach. The colours of the blossoms are gloriously unlike real flowers: it looks as if the place was inspired by the Andrex shelves at Tesco. It's an idealisation of wilderness: sanitised, toned-down, candified. It's the landscape of the Walt Disney film *Fantasia* come to life: as Beethoven's music plays, so from the pond emerges a troupe of naked nymphets; as they continue to walk, they are revealed as saucy little centaurettes, swishing My-Little-Ponytails attached to cute semi-anthropomorphic bottoms. As the golfers in their curious trousers enact their curious rite of spring and their balls slip into the water hazard, I expect the equinymphets to step daintily from the water after retrieving them – and all to music of the glorious Ludwig van. Augusta is as much like the real thing as a glass of Babycham – but for some people it is as sacred a space as any that exists on earth and I am to be both scorned and pitied for my blasphemy.

50

The Suicide Month

October marches on with a sense of mounting horror. Tension has the Valley in a death-grip: a lion's throttle-hold that chokes the life out of you. Every day is hotter than the last. The sun, bringer of life to old cold England, is here a lethal ball of flames. The sky will bring life – you hope – but not like that. In the heart of the heat of the day almost everything that breathes in the Valley is in the shade. Me too. I remember co-leading a trip with Chris, both of us in staff quarters rather than the guest cabins, sharing a hut that stood in direct sun. It was here that Chris invented the three-shower siesta: not just a shower before and a shower after but also a shower in the middle.

Yet some choose to leave the shadows. I watched a male puku step down to the river: the long, long reaches of sand to the thin ditch of the Luangwa. Choosing the worst, the most uncomfortable time of day for the one drink he needed in 24 hours: the idea being that the lions would be too hot to bother just now. Too hot and too full.

October is a time of death: enthralling, beautiful, terrible. The carnivores rule. It is the time of privation for all but them. It's also a time of enthralling, unbearable tension: the world stretched on the rack, like a guitar with the strings tuned just short of snapping point to give a wild, unstable discordant treble scream. It's a grid before the start of a grand prix, it's the shower scene in *Psycho*, it's the seconds before your first kiss – and it goes on for day after day after day: the sky an utterly unforgiving blue, an occasional tiny cloud like an insult, nights in which the temperature seems scarcely to drop. The narrow corridor along the river is thronged with large mammals waiting for death, waiting for rain. Whichever comes soonest.

These times affect humans too. How can they not? They call it the Suicide Month, and for the best of reasons. Even if you don't take that irrevocable step there are times when you can feel a trifle like it. For the residents this is not the trip of a lifetime but the grim routine of the year. And just as our cold Northern hearts sink as the year approaches the winter solstice and the day is a brief and grudging offering of pale grey light, so the impossible claustrophobic inescapable heat brings despair to humans in the Valley. As we English despair in December because the sun will never shine again, so people of the Valley despair in October because it will never stop.

A hippo died in the river just outside Mchenja. The stink was of such an epic nature that we abandoned camp and moved to Kakuli, a few miles downriver. A couple of days later – out on an afternoon drive, the Mosi in the cool box moving nicely towards the boil – we saw a dozen vultures on a tree. And we knew what that meant, all right, but not quite how ghastly it was going to be. A vision of hell awaited us. But first – I was with Bob, after all – we had to identify the vultures: white-backed, hooded, white-headed and near the top a pair of gigantic lappet-faced vultures. Waiting.

51

Climate of the
Luangwa Valley

Temperature in degrees Celsius, rainfall in millimetres.

	High	Low	Rain
January	24	16	170
February	24	16	190
March	25	17	150
April	26	16	45
May	24	14	10
June	23	13	0
July	24	12	0
August	28	16	0
September	32	19	0
October	35	15	10
November	30	15	100
December	28	14	160

52

Tourism

Tourism. Safari industry. Clients. Rules about spotlighting. Rules about walking. It's not the virgin wild, is it? And perhaps that disappoints you a little. Perhaps that even arouses a faint sense of contempt. Perhaps you're muttering to yourself: not exactly a great explorer, are you? Just a tourist. Might as well write a book about San Tropez or Disneyland. And if I have given the impression that I see myself as "the famous Captain Spaulding, the African explorer – did anyone call me schnorrer? Hooray hooray hooray!" – as Groucho Marx sang in *Animal Crackers* – then I apologise.

But sorry, I refuse to apologise for tourism. What, I wonder, would the Valley be without its visitors? Gee whiz, oh my gosh and golly, a lion! A leopard – now I can die happy! Where would the Valley be if it couldn't offer several thousand people the experience of a lifetime every year? There are still many remote unvisited wild places in Africa – and mostly they're poached out: the edible animals gradually killed and eaten, the elephants shot for ivory and the carnivores starved to extinction. The trees felled for fuel. The place uncared for, unloved, slowly ruining. I know, I've been there, in many places across Africa.

Tourists have kept the Valley wild. Had it been left, it would have been tamed by extinction. Tourists bring money to the Valley and they put a high monetary value on the wild places of Zambia – and for that matter, of the world. Not only that, the fact that visitors value the place for something beyond money, means that the people on the ground increasingly do the same. The Valley has become a source of pride for Zambia, rather than one of embarrassment and vague shame. Since I've been coming to the Valley I have seen Mfuwe township grow from a rough spot with a couple of dusty old shops, to a thriving place full of beauty parlours, bars, a first-class school, all kinds of good things. Inevitably this brings problems with higher demand for wood, felled outside the park's boundaries but in the important buffer zones. And despite the presence of game scouts there will also be some poaching going on. But these are infinitely better problems than those that come with abandonment and neglect.

Nor do I despise the tourists. If Norman Carr believed in tourism, then I can hardly go against him. Have you been there? If not, I wish I could beam you up there right now. It would change your life. And certainly it will change the way you see the world. This is a place and an experience that enriches those who come across: why wouldn't I share it? In the heart of the South Luangwa Valley National Park you will see other vehicles, though not hundreds. You won't be sharing a lion with 30 vehicles. More like two or three, on a crowded day. And they'll all behave well: the guides are self-policing and strict with those who step out of line.

A lot of people come to the Valley with the Big Five mentality. The Big Five is a term that has spilled over from hunting: once you had killed the five most fearsome animals in Africa you could consider yourself a real hero. The notion of the Big Five still dominates. It's a need to see the stars: to tick 'em off one by one; to feel you have, as it were, got your money's worth. And I don't blame anyone for wanting this. On my first trip, it seemed for a while that lions would evade us, which made their eventual appearance so wonderful, of course.

But the weird thing is that when you start looking at wildlife – especially when you do so in a place as rich as the Valley – you get more than you bargained for. And people who have never looked at a bird in their lives, start getting excited about birds: they learn their names, they take pride in naming them before the guide, even in learning a call or two. And it spreads out from that: it all becomes interesting. And the guides like Manny are not only able to tell you the scientific name of every tree in the damn place, but they will also have a working knowledge of the multitudinous forms of life in the Valley. And rather more than that as well, mostly from the experience of walking. When I say that a trip to the Valley means you will never see the world in quite the same way again, I am not being fanciful. I'm sticking to the dull, dry, pedantic, plonking, literal truth. Take home that attitude you learned in the bush, and you will look more and see more, and where once you found sterility, you will find life. It starts with the Big Five, but it continues to the Little Five and even to the Plant Big Five.

It was my first visit to the Valley, and I was having a last jar round the campfire at Mchenja with our tour guide. Chris, that was. He was moonlighting: his day job back then was selling aeroplane tickets for Trailfinders. "I want to bring people here," he said. "I want to start my own travel company and specialise in trips to see wildlife." Something about his energetic manner, merry nature and hard-wired optimism had impressed me right from when we met at Heathrow Airport. "If I had any spare cash, I'd invest it in you right away," I told him. "Because you're going to end up doing exactly that." Three years later, the year I made my extended stay in the Valley, he set up his company. His office was the wall-phone in his girlfriend's flat, but that phase didn't last long. A little less than 20 years later the two of us were co-leading a trip to the Valley with Chris's clients from his company Wildlife Worldwide. I wonder how many people have had their lives changed forever by travelling with Chris? Quite a few, but they start with me.

We have to embrace the contradiction: wilderness won't stay wild unless we fill it full of people.

53

Three Lists of Five

The Big Five
Lion
Leopard
Elephant
Rhinoceros
Buffalo

The Little Five
Ant–lion
Leopard tortoise
Elephant shrew
Rhinoceros beetle
Red-billed buffalo weaver

The Plant Big Five
Lion's tail
Leopard orchid
Elephant grass
Rhino thistle
Buffalo beans

54

Uncle Monty's Cat

Disgusting. Appalling. But I'm not disgusted or appalled, not really, because it's the bush and it's the way things are. Certainly you get used to seeing some pretty disgusting and appalling things out in the bush, though that's mostly because you spend a lot of time looking for lions, and lions are capable of some pretty appalling and disgusting things. I mean, they actually eat animals. True, they don't keep them in disgusting and appalling conditions first, but we humans don't have to look at the way our food gets to the dining-place. And lions treat their food with a certain frankness that can be distressing: they don't mix the unmentionable bits of an animal together into a little patty and serve it up in a bun, they just eat the bloody stuff – and when the blood encrusts their faces, a close relative will lick it off with a tongue like a farrier's rasp.

All the same, there really was something a bit dreadful about the dead hippo. It was the size of the bloody thing, for a start. A capsized buffalo looks big enough, in all conscience, but the dead hippo looked like a grounded Zeppelin. So much meat! Even for the core group of the pride, in the inner 12, it was a lot. But for this special occasion, the big male came out of his seclusion and was lording it over the place. Not that there was much competition: there was meat enough for half a dozen prides. But this lot had it to themselves. Even then, I wondered how long the damn thing would last. I mean, meat doesn't exactly keep that well in a hot climate, does it? And now every day the temperature was peaking at around 40 degrees C, 104F. Not ideal for keeping a tasty morsel fresh for a hippo bourguignon the following day.

There was a ghastly air of repletion round the vast carcass. They were all lying down in glorious self-indulgent regret at their own wicked self-punishing greed, and even then the male got to his feet and ambled over to it for a bit more. A cub, hippo gore encrusting his muzzle, padded – wobbled – out to join him. A pride male will often be resentful of too great proximity at the meal-place, but will happily put up with a cub; so long as it's his own, of course. One more mouthful and that little cub would surely explode, but no, more dead hippo please. Absurdly, I was reminded of Uncle Monty and his cat in the film *Withnail and I*: "Get that damned little swine out of here. It's obsessed with its gut. It's like a bloody rugby ball now."

I felt that my brains were sizzling in the desiccated frying-pan of my skull. One of the young male lions got up and, with an air of content, almost of luxury, shat. There was a clatter like the falling of a partly opened umbrella and one of the vultures was there to wolf down the smoking globes. A white-backed, since you ask.

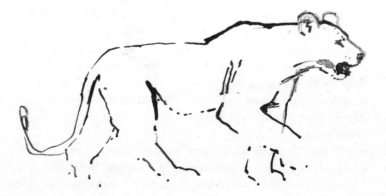

55

Banished from Paradise

I've been writing about threatened wildlife for a long time
and I am used to addressing questions about decline. Why
have numbers fallen so drastically? Numbers of individuals,
numbers of species. Just about every time, it comes down to
habitat destruction. Destroy the wild places of this planet and
you destroy the creatures that live in them. The whole damn
place is getting too crowded: there ain't no more room here,
no more room here, no more room on this planet to grow, as
Grace Slick so justly remarked. I've been in the prairie farms of
East Anglia in England, I've been in the jungles of India, I've
been in rainforest in Borneo and Central America, I've been in
highlands and wetlands and moorlands and all the time it's
been the same story: humans incontinently flowing into every
space. The wild world is like travelling economy on a long-
haul flight when, just before take-off, the most obese man in
the world comes and sits down between you and the aisle. You
can't move, you daren't ask him to move, and he's spilling into
your space because he just can't help it. No doubt it's his
history of greed that makes him that way. All the same, he
apologises to you, he doesn't mean it, he can't help it, but
what's that to you? All you want is room to breathe, and you
ain't got none. His name, by the way, is Mr Creosote: the
Monty Python gourmand persuaded to ever-greater feats of
gluttony by the cringing *Maître d'*, John Cleese. Lions don't
want too much: they only ever want enough. It's that other
species that fancies itself king of the jungle that always
wants more.

But here's a rum thing: Mr Creosote is not present in the
Valley. Habitat destruction is not a problem here. Not yet,
anyway. Right now, the Valley is immune from the vast, the

universal, the overwhelming problem of wildlife conservation across the world. At least inside the borders of the National Park. And that means that the creatures that have always lived there are still living there.

With one exception.

It was only after travelling elsewhere in Africa, notably in Zimbabwe, that I got an eye for the habitat. I was able to see the Valley in the context of all Africa. I have walked through areas of wooded savanna very similar to those in the Valley, filled with trees and bushes of convenient browsing height and with plenty to drink – and in those places I have found black rhino. That's the smaller, faster, fiercer of the two African species: the one with the hooked lip. The white rhino, or wide-lipped rhino, is a grazer that favours open country; the black is a browser that likes thicker bush, using that ingenious and agile upper lip to strip twigs of their leaves – though it should be remembered that rhinos are neither black not white: they are 50 shades of grey and more. Rhinos, great rollers and dust bathers, tend like elephants to be the colour of the country.

The Luangwa Valley should be jumping with black rhinos – and it isn't. They all got shot. For their horns, which are important in Chinese medicine, not as an aphrodisiac but as a treatment for fevers and disorders of the blood. The last one went sometime in the 80s: I was just too late. And so when I walk across the Valley, I see a country that is perfect for black rhinos and where there are none. It's like walking through St James's Park and finding all the ducks gone: while the lake is as inviting as ever and the children are still there with all the bread in the world – everything is perfect, but not a quack to be heard. It's a spooky business: a bit like the time in *The Hitchhiker's Guide to the Galaxy* series when all the dolphins – being hyper-intelligent beings – abandon the earth. Hence the title of the fourth volume: *So Long and Thanks for All the Fish*.

The rhinos are gone. Gone to provide a cure for stricken people, except that the cure doesn't even work. The Luangwa

Valley is a paradise for black rhinos. But it's not the paradise that's been lost, it's the damn rhinos. You can make a perfect place in your own mind, but if you seek such a place in the living world, you will find flaws. No matter how magical the place and no matter how dazzled you are by its perfection, you will find that perfection is not, after all, an absolute thing.

56

Nothing Wasted

We visited the late hippo every day, and every day it was worse. There was the day all the maggots erupted, and yet the hyenas, who had by then taken over from the lions, saw them only as more protein – well, they'd be fools to see it as anything else. There was a vast slab of moving meat and a horde of grinning feasters. For two days they waited for the lions to finish, occasionally making a mad giggling run when the carcass was untenanted, to be chased off again, sometimes with the prize of a slavering gobful of rank meat. But even lions can have enough. Eventually they were too full to bother about safeguarding what was left – and anyway, there'd be another meal along in a moment, for the time of plenty was not yet past. After a while, even the hyenas were sated and it was the time of the patient vultures. They looked like a hundred jabberwocks with eyes of flame, whiffling across the tulgey savanna, hissing and devouring. Sometimes one would get so full it was unable to take off. Even those that managed this difficult feat made it only to the tree tops where they sat and digested until they were able to resume the glorious gliding life again. They are birds of astonishing beauty of movement, but their way of life appals. And all that bare skin around the face and neck: it's a fine adaptation if you earn your living by sticking your head in a corpse – you don't want your feathers all clogged up with blood after all – but it's not a winning look in human eyes.

Eventually even the vultures had gone, and so had most of the hippo. I remember a vision of black bones and a little black flesh. I was reminded of the Tom Lehrer song about the girl who drowned her brother: "All they ever found were some bones – and occasional pieces of skin". Alongside, just as black, three ground hornbills, the size of turkeys but with beaks like meat cleavers, helped themselves to the last scraps. And one final visit, a hyena had come back to lick the bones like a lolly. He was lying within the carcass, so he was looking out through the bars of a cage. A grin of idiot satisfaction on his face.

October is the cruellest month; the last week cruellest of all. But not if you're a carnivore.

57

Pretty Woman

A trip to the dead hippo could be varied with a visit to the carmines, for they are at their most spectacular at the end of the dry season. When the heat gets personal, the old hands talk of suicide, the carnivores look ready to explode from overeating and the antelopes look as if they'll burst into tears at the least provocation, so the carmine bee-eaters seek to redefine the possibilities of beauty on the parched, naked sand-banks that line the great shrunken river.

One carmine alone is the stuff of dreams. Look at a picture and you will at once agree that the bird is impossible. No bird could be quite that wonderful, and even if it was, it's obvious that you could never hope to see one: too remote, too lovely, forever beyond your scope. You know from the picture of a carmine that you will never deserve such a bird. No one could. It's a bird too far: made of raspberry pinks and Burgundy reds set off with million-volt electric blue on the head and the bum. It has a melodramatic black mask to set all this off. The shape is streamlined and elegant, and the tail has a long central streamer. This is the bird that paints the lily, that gilds refined gold; that improves its own perfection with still more perfection until an onlooker begs for mercy. A single bird is a masterpiece.

But you never see a single bird. Walk along the banks of the river and you will find a colony: a thousand birds together. It's like an explosion on Murano, the place where they make Venetian glass. It's as if someone had raided the island's rubbish-tip and gathered tons and tons of shards and splinters of the brightest chandeliers, candelabra, tropical fish, cockerels, mermaids, horses, temple-maidens, clowns and loving-cups and hurled them into the air for the sun to turn them, at a stroke, from kitsch into perfect beauty.

Carmine bee-eaters nest in burrows in the friable sandbanks. At the end of the dry season they are suddenly hell-bent on improving them: digging deeper, repairing collapses, expanding brooding chambers. You constantly see little puffs emerging from the entrance, ejected heaps of fresh-dug sand: it's like watching a cannonade from a man o'war. The birds fill the air above the colony, flying in tight swooping curves, hawking for insects or sometimes, it seems, just going round and round to celebrate the joys of togetherness, of being part of a colony, all setting off together on the great adventure of making more carmine bee-eaters. Their voice, it's almost a relief to report, is not lovely. A party of carmines sounds like a flight of Yorkshire terriers passing overhead.

It's best to observe them from the opposite bank, when their frenzied comings and goings eventually make a kind of pattern. I wonder how they make their choice of mate, for both sexes are equally and indeed identically beautiful. To a human at least, they all look the same. For a male carmine it must be as if all the females were Julia Roberts, and all the males too, himself included. Sometimes you see a stray party of humans from another camp on the other side of the river. As they step from their vehicle the bee-eaters leave their burrows in one synchronised moment of alarming glory, and everything in the world is cherry-red and million-volt blue.

And meanwhile the river dwindles beneath them and the sun continues to burn its unforgiving unforgivable arc across the relentless blue of the sky.

58

Rain

We paid a visit to Derek who, that season, was setting up Kayango Camp a little upstream of us. It was in a fine location, but then there are only fine locations in the Valley. It was an exciting job they had taken on, for this was not an established camp busy with humanity and light and fire. It was just two men in the bush, and the large mammals they shared the place with didn't make much adjustment for them.

We sat on logs at the top of the bank, for the place had yet to acquire such decadent luxuries as chairs, and we had tea. And that's when the strange thing happened. In the length of the time it takes to make and drink a cup of tea, the season changed. First, clouds came hurrying into the sky as if late for an appointment: not single puffy white things but a great bank of blackness. It was as if a vast iron curtain were being drawn across the sky, with no Checkpoint Charlie where an errant bit of blue could get through. A wind struck up from nowhere, making me wonder if there hadn't been a breath of wind here for weeks. And it was cold, as if someone had opened a window in a centrally heated house. For the first time in two months, I felt the need of a garment to cover my cool-in-the-bush green shirt. It felt like the cold breath of England.

Then all at once, far above I heard a deeply familiar and profoundly strange call. European swifts! Impossible but true, high, high above, so high it was only just possible to make out their fine sickle-shapes. It was the call I had heard a million times above the cities of England, above Wimbledon when I covered the tennis on Centre Court. And here was, perhaps, the most amazing rain-delay I had ever witnessed. Lightning made startling patterns across the sky and thunder rumbled on and on like the hundred-letter — more properly thundred-letter — words in

Finnegans Wake. James Joyce was mortally afraid of thunder, but he never visited the Valley. Had he done so, he would have known that in some places, thunder is not a curse but a blessing.

And then a damp gust bringing rain.

Each drop was heavy as lead. In a few moments the sand before us was pockmarked, as if hit by a blast from a giant shotgun, and my shirt was so spotty I had become a cool-in-the-bush Dalmatian. The flat surface of the river was agitated into a million little puddles.

A miracle. I had witnessed a miracle.

It was time to go home.

59

A New Brain

It was good to be home: better than good. I missed Africa, but not in the sort of way that means I wished I was still there. I really didn't. Not that this was entirely obvious to an observer. Cindy remembers me on my return: a silent figure with a thousand yard stare. "It was as if you'd left your lover," she said. Even so, I was where I wanted to be and deeply happy about it. Even if it didn't show. I just hoped that I'd be back in the Valley pretty soon. And besides, what I found in Africa can be found elsewhere. In Africa, I experienced the wild world as I had never known it before, but that wasn't something that got lost as I got on the plane at Mfuwe airport. Something of Luangwa was with me even in cities, and always would be. Sure, I was a half-decent birder before I went to the Valley for that extended stay, but after coming back I was different – but not because I was always mooning after Africa. No; after that trip I was connected as never before with the wild world. For two months looking and listening had been the most important activities of my life and I have never really lost that, no matter where I happened to find myself subsequently. My senses were no sharper than they were before, but I had trained my brain to prioritise information about non-human life. There are some experiences that literally alter the chemistry of your brain – perhaps that's what had happened. Perhaps I had realigned the neural pathways. Because it wasn't quite the same brain I left England with. I found that my eyes followed the flight of a city pigeon as readily as I had followed the flight of a martial eagle in the Valley. There was no decision to make, no sense of volition: my eyes did it of their own accord. In the same way my ears picked up the sound of a great tit in a suburban garden when I walked to the pub as smoothly as I had picked up the calls of orange-breasted bush-shrike and water dikkop when walking out in search of lions.

The wild world and the wild creatures that live in it were no longer a bonus. Wildlife was no longer an add-on, something that I could go and look for when I had a free weekend. It was no longer something I could pick up and put down. The wild world was with me always: on train stations, in pub gardens, at the cricket. Once, while batting, I stopped a fast bowler in mid-run-up to point out that a cuckoo was flying overhead; it was perhaps fortunate that we were playing the RSPB at the time. The bowler and all the close fielders – as usual when I was batting, there were plenty – followed my gaze and complimented me on a good spot before going back to the cricket.

In a sense, I was now always in the Valley. As an alcoholic always knows where the next drink is, being hyper-aware on that subject, with a brain trained for the one thing that really matters to him, so I always know where the birds are. I've come as close as I can, not just to finding the paradise island in the song, but to finding it everywhere. Sometimes it's as joyful a thing as the song suggests, but it would be wrong to say that my life is now a progression from one joy to the next. After all, does great literature always make its readers joyful? No indeed. Dr Johnson said that the job of a writer was to show his readers how better to enjoy life, or how better to endure it. The wild world can bring joy, and can bring solace in bad times. And both those things are fine and wonderful, but that's still not the whole story. If you have schooled your brain to pay attention to the wild world, you come to some kind of understanding. Life changes on you. You see everything in a slightly different way, and what happens is that wild life and tame life become indistinguishable. You become part of a continuum. Does that make life better? I think it probably does, but I don't really care if it doesn't, because now, I wouldn't want to live in any other way. I left a piece of my heart in Luangwa, but Luangwa left a piece of itself in me. It seems to me a fair swap.

60

Fight or Flight?

I was approaching the top of Mount Kinabalu when I made a resolution. No more bungee jumping! A great, murky, misty dawn was breaking. Well, sort of; I was inside a cloud at the time so it was hard to tell. I was breathing with the wheezy resolution of asthmatics everywhere. I was going to reach the top, I was going to get a certificate to prove it and then I was going to come down again. So bloody what? In two days of walking up a hill there had been far too little time to appreciate the place, to look at the wildlife. I had hoped to see the Kinabalu friendly warbler – so called because it's endemic to Kinabalu National Park in Borneo and, in the pioneering days, was amiable to the point of being downright confiding – but alas, no. There wasn't much birding opportunity. It was all going up and then going down again: reaching a peak that is 13,435 feet / 4,095 metres above sea level. So it was just a stunt. All I got from doing it was the pleasure of saying I'd done it. Which wasn't much, to tell you the truth. So that was it, I decided. No more stunts. I'd do any amount of crazy stuff if there was a point to it, but if there was no point, no thanks.

A few months later John Coppinger asked if I'd like a trip in his micro-light. We were at his lovely camp in the Valley, Tafika, which means in Nyanja, "We have arrived".

I knew what that entailed: riding pillion sitting astride a thing like a Vespa, only much less rugged, with a kite over our heads and a lawn-mower engine between our knees. So I said I'd think about it. I did, too. I wondered: was this another stunt? Another bit of bungee jumping? It didn't take me long to realise that it was nothing of the kind. It would be seeing the beloved country as I'd never seen it before. You bet I was up for it.

I was scared all right. I was scared during the safety briefing; scared trying to sit comfortably; scared as we trundled onto the brief runway which is cleared regularly by panga, the local form of machete. I was very scared indeed as we roared (rather asthmatically) towards the take-off point, the little engine farting hysterically beneath us. But I stopped being scared the instant we left the ground. In less than a second the great mad snake of the Luangwa River was lashing away beneath us, threshing from side to side. There were the great wide sand beaches; there were the bends so extreme that they almost met in the middle, and there were the oxbow lagoons where this process had been completed. There were the darker crescent shapes of old lagoons – a different colour because the vegetation was different – that had silted up and died, returning to pure bush even as new lagoons sprang into being: a process that is eternally self-renewing. On either side of the river, the great flat valley-floor spread out like a ballroom, one in which the eternal dance of life and death had been taking place from one millennium to another.

The land spread out beneath us to the distant walls of the Fuck Me Hills and the Fuck Off Escarpment and, running right through it, was the unravelling skein of water, created as if a straight line were a kind of blasphemy. So much water. I hadn't imagined it was possible to place so many sorts of water in one place: river, tributaries, tributaries to tributaries, streams, lagoons, lakes, pools, ponds, puddles. As the swallows and swifts that love to gather around, above and below the Luangwa Bridge fly only in endless curves, so the river beneath them sticks to the same plan. The Luangwa River is ever-changing yet always the same, and it made sense of everything else I could see. Here was a river designed entirely by the forces that make rivers, for the hands of humans have scarcely touched it. It was a river as wild and untamed and uncontrolled as any of the great mammals that dwell alongside it and in it. The river is the heart and soul of any valley, spinal cord, blood, principal arteries and all. We climbed high amid the early birds of prey and then came down and made a kind of toboggan run between the banks of the Luangwa River, shaping our flight to its irrational curves and convolutions. I had always understood the river in my mind: now I understood it deeper than did ever plummet thought.

61

Crocodile Fears

There are no alligators in the Valley, of course, not even white mutants: only two species of alligators survive in the modern world, the American and the Chinese – distinguished from other crocodilians by the breadth of their snout. But the Valley might be the world's crocodile capital. These are Nile crocodiles: the world's second largest living reptile, after the saltwater crocodile. They are intensely social, but that doesn't necessarily come across as an endearing trait. In the dry season they gather together, one hundred at a time, in deep pools in what's left of the river. Shine a light on them at night and the water lights up like the fountain in Las Vegas. The daytime sight of a river paved with crocs is disturbing to the human sense of what's right. It looks as if you have returned to the late Cretaceous, and why not? Crocodilians lived alongside the dinosaurs towards the end of their 100 million years of dominance, and it was the crocs that had what it takes to survive the meteor strike that did for their monster fellow-reptiles.

A croc lives and dies and defines itself by means of its bite. Its life is based around sharp conical teeth and a grip that can't be broken, one they can hold for extended periods of time, so that any land creature grabbed must surely drown. I have waded the Luangwa many times, once armpit deep, and it was never something I felt comfortable with. Crocs take fishermen regularly: it's a perilous occupation, even if it's intermittently rewarding. There are places along the river in the dry season where a gathering of crocs shares the same deep river-pool with a hundred or so drought-stressed hippos. Aged and diseased hippos die in these times of weakness, their destiny just a few yards away. The crocs move in and share the giant meal: seizing an epic mouthful and then performing a full body-twist to rip a morsel free, the pale underbellies of the crocs briefly flashing skywards.

I have canoed on the Luangwa, again never entirely at my ease, but then you're not really supposed to have an ease in such circumstances. I once saw a croc take a puku antelope, or rather I didn't see it. The crocs are that quick: one moment the puku was drinking, the next it was drowning, four dainty cloven-hooved paws pointing to the indifferent sky. On another occasion, I was there as a croc took a baby elephant.

They are, then, impressive beasts. John Coppinger once canoed the length of the Luangwa and the experience gave him a deep wariness of crocs. One afternoon he was returning to his camp, Tafika, in his RIB – rigid inflatable boat – laden with stores, for this was the wet season and the river was high. A croc appeared on the surface ahead of him and gave him what seemed to be a nasty look. Well, John reasoned, if you advance on a lion, it will generally back down and retreat, so presumably a croc will do the same. So he steered straight at the damn thing. It opened its mouth without hesitation and ripped a huge gobful from the boat. The croc considered the matter closed at that point, which was fortunate. John had to cover the next 20 miles as the boat deflated, for the tube was not constructed in discrete airtight sections but was all of a piece. Before long the boat was tugging a vast underwater skirt. It must have looked like the Victorian lady who attempted suicide from Clifton Suspension Bridge (near where I didn't buy the Manhattan White) but was sustained in the air by her crinoline, or rather, the skirts it supported. She was pulled from the river surrounded by immense acres of sodden skirt. The outboard motor was somehow able to cope with this terrific resistance, but it wasn't a quick journey. He made it all right, and no doubt by the end he was giggling like a schoolgirl. And he had another campfire tale.

62

Are there Werewolves Still for Tea?

Paradise can exist in time as well as space and in both cases, of course, it is largely imaginary. The Golden Age is a moving fixture in human consciousness: always about one and a half ages before our own. The years before the First World War are regarded as an idyll: a time of privileged perfection. This was also – after Eve's proffering of the apple to Adam – one of the greatest *look-behindjer* moments in all history. Rupert Brooke, scribbling homesick lines in a German café and, in the second verse, not troubling to hide his antipathy for the German Jews drinking beer at a nearby table – no one quotes that bit for some reason – summed up forever our notion of the Edwardian idyll in the poem's concluding couplet:

Stands the church clock at ten to three?
And is there honey still for tea?

Time itself has stopped. Nothing will ever change. We are locked in an endless era of peace and prosperity with a comfortable unassuming dominance over the entire world. And we 21st century humans can look back in glorious vicarious nostalgia and savour the fact that this paradise was built on the slopes of Vesuvius. The writer who best captures these times is Saki, pen name of H. H. Munro, who died in the First World War; his last words were: "Put that bloody cigarette out." Before the war he was the great narrator of Edwardian drawing-rooms. His best stories are fantasies of the wild world's invasion of this false paradise, hinting at the real passions and horrors that lurk in every human breast and are all the more vivid for their backdrop of chintz, cucumber sandwiches and visits from the vicar. My favourite tale is *Gabriel-Ernest*. It tells of how the highly respectable Van Cheele finds a naked boy sunning himself in the woods. "At night I hunt on four feet," says the boy. Van Cheele tells him to leave – so next day the boy then turns up, just as naked, in Van Cheele's house. Miss Van Cheele, the owner's aunt, takes pity on him and gives him the name of Gabriel-Ernest. "Clothed, clean and groomed, the boy lost none of his uncanniness." The pet spaniel bolts, and lurks shivering outside the house. Meanwhile the aunt arranges for Gabriel-Ernest to help out with the little children at the Sunday school tea while Van Cheele visits an artist friend who had seen something odd in the woods: a boy, naked in the setting sun. The sun disappeared: "And at the same moment an astounding thing happened – the boy vanished too!" What, asks Van Cheele? Altogether? "No; that is the dreadful part of it… on the open hillside where the boy had been standing a second ago, stood a large…"

But come. Read the rest for yourself, in Saki's own words. Just Google Gabriel-Ernest. But before you do so, let's savour that moment when Van Cheele finds the boy in the morning-room. "Gracefully asprawl on the ottoman, in an attitude of almost exaggerated repose, was the boy of the woods." Naked, insolent, sinister: a threat to the whole safe world. Saki wasn't destroying the idyll, he was improving it. It was the illusion he was destroying. Oh yes, you always need a bit of edge to give meaning to an idyll.

63

Death and All That

Her name was Lyn, a gentle-mannered lady of a certain age, witty, good company and taking every delight in the Valley. I was co-leading a trip with my old friend Chris, and Lyn was one of six clients he and I were taking on behalf of Chris's company Wildlife Worldwide. We were doing our night-drives in two vehicles; Chris and I were taking turns as to which group should have the questionable benefits of our experience. This time Lyn was with Chris, along with Hilly and Hazel, who were about the same age as Lyn. Hazel had had very little experience of wildlife and none of Africa, so naturally she was madly in love with everything she saw. The Valley had her good and proper. On one occasion she expressed her feelings to Chris: "It's just so… it's just so… it's all so… ohhhh!" Hilly, an experienced observer of butterflies and moths, was also delighted with the place, with the depth that greater knowledge gives you, but without the sense of discovery of a completely new world, a completely new way of being.

And it was one of those evenings when those of us who know the wildlife of the Valley can smell trouble in the air. By this time, I was among that number. The rains were expected almost hourly, and yet every morning the sun came up one degree hotter than the day before and the cruel blue of the sky arched above in threat. Then night fell and it was scarcely cooler and much more dangerous. These are thrilling yet deeply troubling times. This time of year is like being at a football match a few minutes before someone gets sent off: a wild and oppressive atmosphere in which you're waiting for something to burst, some radical change to take place, some kind of violence to happen before your eyes.

We had driven to a lovely group of ebonies, using the spotlight to seek out the hunters of the night. This was a place that Manny, having a romantic soul, called the Robin Hood Glade, and we found a very nice group of lions, nearer to us than to Lyn and Chris's vehicle. There was an adult female and a sub-adult female, along with three cubs, enjoying each other's company, the fubsy cubs playing in a rather hot, desultory fashion. We were taking a good look at these contented and peaceful lions when a herd of about 100 buffalo muscled their way in, softly lowing and grunting to each other. It was like dramatic irony: we had privileged information denied to those on stage, but unlike a pantomime audience, it wouldn't help to shout "Look *behindjer*!" Perhaps it would have been unethical. Because the buffaloes knew nothing of the lions that were now in the middle of the herd. I could see that the big female lion was agog to hunt, but she needed help to take on a fully grown buff. Besides, it was hard to pick out a single individual in this seething melee. The air was full of movement, crackling grass underfoot, the snapping and swishing of the bushes that got in the buffaloes' way, the soft thud of hooves, the click of horn on horn. The air was full of dust stirred up by the buffaloes' passage, tickling the nose, rasping the throat. You couldn't see very much. It was all enigmatic shapes and movement.

Then bang. Crash. A scream. A horror. A clamour and a desperation almost on top of the second vehicle: two huge male lions had risen as if belched from the guts of the earth and, before anyone human or bovine could register their presence, they had knocked down a buffalo calf. Snarling, squealing, threshing, struggling: panic among the rest. All the buffs knew there was trouble, but few knew exactly where. Some fled straight *at* the male lions, others straight at the female and cubs. It was then that the two males had a falling-out. They turned on each other and stood on their hind legs and grappled while making a sound from the bottom of hell. The question of precedence, it seems, takes precedence over the question of food. So the calf wobbled optimistically to its feet and made as if to go galumphing off – whereupon one of the males broke off combat and hauled it back. No. Sorry, little thing, but you're staying. So then, dispute forgotten or postponed, one of the males despatched the calf – it was four feet high at the shoulder – with the slow, powerful smother-grip that lions use. There was chaos and dismay in Chris's vehicle too. Hazel was in tears, Lyn pretty close. Chris said that's it, no problem, let's go, let's pull out right now, let's leave them to it. "No, no, we can't go, we've got to stay."

In milling frenzies the buffaloes withdrew, leaving just one of their number behind. It seemed to me they had got off lightly. Through the dust-filled air I could hear the sounds of butchery, horribly graphic. It was then that Lyn announced: "I'll put my fingers in my ears so I won't see."

64

Blood etc.

You expect the conversation at a gathering of sports journos to be fairly robust. We were talking about theft, and so I contributed a classic example of kleptoparasitism – that is to say, the theft of food between animals. I told them how I had seen a leopard stalk and kill an impala, and how a hyena had stolen the kill from the leopard, and how we had – wrongly, on a clientless night – intervened and driven the vehicle at the hyena to chase it off, and how the hyena had indeed retreated, but not before snatching at the impala's belly with jaws filled to overflowing with teeth. We saw it running off into the night with an impala foetus hanging with horrible limpness from its jaws.

My colleagues were revolted by this. After an awkward pause, the subject was changed. They went back to their discussion about mugging and other forms of robbery with violence, and people they knew who had suffered from such things. And I was surprised; I am still surprised. My colleagues tend to think I'm a bit of a wimp, being a birdwatching horse-riding vegetarian. But this tale of everyday life in the bush upset them hugely. On another occasion, I remember describing to my wife – by no means lingeringly – a fairly seismic encounter with predators. And she said dismissively: "You like that sort of thing."

It's an idea that troubles me. I love seeing great predators and watching them go about their terrible business, business for which they have evolved so gloriously. I don't like seeing beautiful animals like impalas being slaughtered; I'm certain that if it were a human doing the slaughtering I would put my fingers in my ears and do anything else it took to prevent my seeing it. It's different when it's a leopard. And when the two lions killed the buffalo calf, I had no need to look away. It was terrible, but it was also wonderful. It's how the bush works; it's how the Valley works; it's how ecology works.

The lesson, I think, is about the difference between human morality and the way the wild world operates. It's as well not to impose either one on the other. Human society doesn't work in a dog-eat-dog way, as a certain type of bore will sometimes claim. Human society is more inclined to work on principles like affection, family, forgiveness, reconciliation and small mutual favours. It's not all about the strong preying on the weak. Humans can do some pretty terrible things to each other, but not because we are obliged to by evolution and ecology.

I have often seen gentle people reduced to tears by the violence of predator on prey. And here's a rum thing: none of them has been vegetarian. I am reminded of Uncle Monty in the film *Withnail and I*: "I can never touch meat until it's cooked. As a youth I used to weep in butcher's shops."

A Manly Tear

Though as a matter of fact I did weep, that night of the slaying of the fatted buffalo calf, though later. Every now and then the Valley goes raving mad and throws one improbably wonderful thing after another at you. You often have nights in which you see nothing but a few genets and a porcupine (and what's wrong with that?) but this was the other kind of night. Still with hearts and minds most profoundly stirred by the lions, we came across a leopard, hunting alone, a female – small, neat and gorgeous. We had her in sight for just a few minutes, moving as if every joint had been bathed in a gallon of oil and glowing as if lit from within. That's one of the nice illusions you can get on a night-hunt in the Valley: the bright light from the spot seems to come back from the leopard still brighter. On the way back from this most stupendous night the temperature dropped like a brick and thunder dryly crackled. And then, grudgingly, a little rain fell. Just a hint, just a taste, just a hint of a promise, but a sign that the times of ease and plenty would not be much longer delayed. And a drop or two that was not rain touched my cheek.

The Unenchanted Combe

I have seen the magic arrive. I've been in places that have nothing particularly special about them and witnessed their transformation. I have seen a combe putting on its enchantment like a magic cloak. Sweden, evening, spring. Not far from Lake Hornborga. Me, dressed for an English winter and feeling the need of another layer at least, but standing there in any case, dancing the stately dance of the cold: slow, stamping, rhythmic shoulder clapping, gloved hands in and out of pockets. And in the middle of this unenchanted combe, just me and Chris. The sun was heading downwards, just about, already showing signs of the reluctance that will come close to a complete refusal to vanish in a couple of months. And, as is usual in such circumstances, the true lover of nature wonders why on earth he ever bothers with such stuff. Cold, tired, there's a bar full of beer not far away and anyway it's obvious they're not going to come.

But then they did. Of course they did. The sound of the wings, the bugling calls from one to the other, the landing of an airforce of cranes. So let's have some Homer, translated by my friend Jeremy Mynott and quoted in his forthcoming book *Birds in the Ancient World*. It's the first reference to birds in the classics.

> *Just as the many tribes of winged birds –*
> *geese or cranes or long-necked swans – gather*
> *on the Asian meadows by the streams of Caystrius,*
> *flying this way and that, exulting in their wingbeats*
> *and settling ever further on with clangorous cries*

until the meadows resound with their calls.
Just so the many tribes of men poured forth
from their ships and huts into Scamander's plain,
and the earth echoed with the dread sound
of their pounding feet and horses' hooves.

The cranes flew directly overhead, exulting in their wingbeats, and they landed about half a mile beyond us with clangorous cries. They stopped there, almost as tall as a human and with a lanky two-legged stance that makes them look a little bit more like us than other birds. They are so completely graceful it seems absurd to use the same word for any human movement. They touched down with the same reluctance of the descending sun, and when they did so it seems as if the land ever-so-slightly rejected them: as if they must bounce back a little, like the opposing poles of a magnet – not north and south but the eternal opposition of ground and sky. Even when cranes stand still they seem to be suspended above the surface of the earth by a distance of, say, half a centimetre. Fenchurch, the beautiful heroine of *So Long and Thanks for All the Fish*, was afflicted with the same thing: her feet were minutely but irrevocably unable to touch the ground.

Even as the cranes settled for the night, they fidgeted and shifted position, and did so by bouncing: by rising and falling a few times. It seemed absurd that creatures so large should be so buoyant. All the time they muttered to each other, in a language that was a strange mixture of muted bugles and purrs. Every so often one would spread its wings like a cloak, as if flamboyantly adjusting it, matador-style, to maximise its contribution to a warm and comfortable night. And so we left the newly enchanted combe – which would become unenchanted once again shortly after dawn – for the enchanted bar, even though we were already high as migrating cranes.

This was a routine refuelling stop on the great journey. The cranes stop here for about a fortnight every year, to feed up, regain condition after the long flight and get into the best possible shape for the breeding grounds. The feeding grounds were at Lake Hornborga itself, a few miles away from the roosting ground where we saw them land; we visited them there as well and saw them hard at it. They would soon fly on to the wilder lands in the north to breed. Here they were betwixt and between: the lust of the journey was still dominant in them, but even as they prepared for the last leg of the great biannual voyage, the lust of mere lust was beginning to affect them. So they danced. What else would anyone do? When they got to the breeding-grounds they would dance in earnest, a vital part of togetherness, bonding, courtship and copulation. But here dancing was more like an expression of promise, of hope, of impatience. It was like watching children coming downstairs early on Christmas morning to stare at the unopenable parcels beneath the tree, knowing that soon all that they could desire would be theirs. And so one bird would unfurl his cloak of wings and dance the buoyant dance of the crane: a hop and kick and a matador-swipe with the wings; not the full thing, but a hint of what was to come. The dance would spread across the flock like a whispered message. Sometimes at the end of a football season the club's fate depends on results elsewhere, and people in the crowd will be listening to radios (if they're traditionalists) or checking on their phones. And a ripple of elation will spread through the gathering as a goal is scored by the right distant team: a widening ripple of joy. In the same way, a ripple of elation would spread though the feeding cranes, and each one in turn, to the rhythm of a Mexican wave, would dance a few steps. And if you happened to be a mere human, you wondered if anything lovelier had ever existed on the face of the earth.

There are 15 species of cranes across the world. One of them was once common in England, but they got hammered in the Middle Ages, largely because, it is assumed, their presence was required at too many medieval banquets. "Let a crane bleed in the mouth, as thou didst a swan, fold up his legs, cut off his wings at the joint next the body, draw him, wind the neck about the spit; put the bill in his breast: his sauce is to be minced with powder of ginger, vinegar and mustard." This is quoted in *Birds Britannica*, the indispensable work by Mark Cocker and Richard Mabey. The finishing touch to the population of British cranes came with the draining of the fens.

Every species of crane has magical associations for the people who share places with them. They seem to be birds that cannot fail to have deep meanings for humans. Many species migrate, arriving and departing as if by magic, sometimes to bring the spring with them – as here in Sweden – sometimes to provide comfort and hope in the leaner and darker months. All species dance; they are all incomparably graceful, and in every culture they pop up in they are seen not as creatures from the lower orders but as higher beings sent down for our edification. The Chinese call them the birds of heaven, and Peter Matthiessen used that for the title of his book about cranes. With calm, confident hands you can make a crane from paper: an origami bird that has a special meaning. If you make 1,000 of them, a real crane will grant you a wish. Sadako Sasaki resolved to fold her own thousand-strong paper flock before she died, knowing that her death was imminent. She died of leukaemia at the age of 12, as a result of the bomb that fell on Hiroshima, and now the folded crane is a potent part of the meaning of that terrible event. I went to Hiroshima and I wept. Everyone who goes there weeps. There is no alternative. I wept not just for her death and for the many deaths and for the madness that humans have created over and over again. Also and especially, I wept for the hope. The hope expressed by the folded cranes.

It seems to me that cranes are not only an emblem of the hope that exists in every human heart, but also, in some ways, the cause of it. If ever you seek hope, the best direction to look is skywards: to see it in all the birds, all the creatures who are excused gravity just as army malingerers are excused boots. All birds can bring just a little hope with them – Emily Dickinson said that hope was "the thing with feathers" – but cranes can bring hope in large dosage. They do this because of their beauty and their grace, but also because there is something human in their silhouette, in their walk, in their love of dance, in their dance of love. We would be like that too, if only we were angels instead of humans. As I left the roosting cranes – me: muffled, cold and ungainly; them: full of grace – I knew that in secret, my heart was dancing as gracefully as any crane that ever took a step.

Cranes of the World

Black crowned crane	Sub-Saharan Africa
Black-necked crane	China and Ladakh, India
Blue crane	South Africa
Brolga crane	Northern Australia
Demoiselle crane	East Africa, Central Asia, India
Eurasian crane	Northern Europe, Northern Asia. Wintering grounds in Spain and North Africa
Grey crowned crane	Southern Africa, including Zambia
Hooded crane	Russia, Northern China. 80 per cent of the world population winter in the Japanese island of Kyushu
Red-crowned crane	Japan, Korea, China
Sandhill crane	North America; the world's most numerous species
Sarus crane	India, Nepal
Siberian crane	Northeastern Siberia, wintering along the Yangtze
Wattled crane	Sub-Saharan Africa; 50 per cent of the world population in Zambia
White-naped crane	Mongolia, Northeastern China, Russia. Wintering in the Yangtze Valley
Whooping crane	Northwestern Canada and Wisconsin; wintering along Gulf coast of Texas and the Southeastern United States

Crowning Glory

How I love the spell-checker. I tend to type very fast when the fit is on me, and very, very inaccurately. The result is literally illegible to anyone in the world who is not me. Sometimes it takes me almost as long to spell-check as it does to write… and sometimes I wonder what it would be like if I used the spell-checker blind. There would be a touch of William Burroughs in this: the intrusion of chance into literature. Or to be more accurate, the knowing, deliberate and self-conscious use of chance. So, what if I accepted every first choice the device came up with? You know best, computer, tell me what I really mean. The result might be absurd, but it might be beautiful, who knows? It might do the job much better than I can.

So there I was, spell-checking the above list of world cranes, and my spell-checker offered this for sandhill crane: "the world's most numinous species". I had intended to type "numerous" of course, and corrected appropriately, as you see, but I wondered for a moment about letting it stand. Cranes are perhaps the most numinous group of birds in the world, even though I wouldn't put the sandhill at the very top of that list. In terms of pure spiritual exaltation, I'd be inclined to favour the Eurasian crane, but I've only had personal experience of five species. Some people make a life's work of seeing all 15: Chris once took a client to see the wattled crane in Zambia, and when they had reached their goal, the client, male, shed copious unashamed tears. As you should, on the whole. I remember sitting in a boat in the middle of Xigera lagoon in the middle of the Okavango Delta as a vast sun made its hurried – no reluctance about tropical suns – descent, when three wattled cranes flew past us. No, I said, please. Not against the sun. That would be too much. And they flew their stately way, black on blood-orange, perfect silhouettes, wattles a-dangle, and me down below quite dazzled with a sacramental bottle of Hansa beer in my hand. Pretty numinous, I think you'll agree.

But it's the crowned cranes that dominate the Valley – the grey crowned crane to be more precise – each bird, male and female, wearing the regal headdress their name suggests. You find them in gatherings of a hundred and more, sometimes many more, in places where the ground opens up and there aren't so many trees: so always you stop and watch them. And always they drift away from you: not a retreat, just a slow but inevitable withdrawal, as if propelled softly but inevitably by the gentlest breeze and, as they do so, they tell you that no one has ever had a sacred combe on his own terms. The crowned cranes let you know, politely but clearly, that at bottom you're not to be trusted, that you can only ever be a trespasser. They tell you of the beautiful frustrations of the wild world. When you actively seek the great experiences, you find yourself clutching at the soap in the shower: the harder your grasp the more rapidly the prize slips from your fingers.

The crowned cranes are bold yet shy – not panicking at your arrival, but fading away. You thought there were a couple of hundred but no, there were only fifty, a dozen, none… And they talk to each other in the bugling tones of cranekind, this species mostly doing so in two syllables. When they take to the air, they sound the bugle on a grander scale, two notes separated by a sort of dirty octave: o-*wan*…o-*wan*. In Nyanja they are owani, and colloquially in the Valley they are owani to everyone.

69

All One

God made the crowned cranes as birds of peace, and their job is to warn us about the dangers of this world. You can tell this from their beauty and their far-carrying voices. They remind us constantly that in nature we're all in it together: united by fear. A walk in the bush will confirm that: for the lovely Valley is a fearful place for humans and for everything else that lives there. *Nimvela owa!* In Nyanja that's "I am feeling fear" – "I'm getting the fear", as they say in the film *Withnail and I*, quoting William Burroughs. Now you should understand that to an African, language is not an insuperable barrier between people. No: it's an elastic concept that unites; Africans tend to slide between one language and another without troubling to consider the matter. Manny was the best Italian speaker in the Valley because he picked it up from his Italian clients (and to this day addresses me as *dottore mio*). Thomas, one of the scouts, would speak nothing but Italian after a few beers, to the bewilderment of his companions who worked in camps without an Italian clientele.

So when the British came to Zambia and English became the official language, English words trickled into the culture, touching all kinds of profound and ancient notions. As a result of this subtle process, the cranes began to call *Olwan… olwan… olwan!* All one, all one! All as one: all of us creatures sharing the same world, united by the same fear, united by the fragility of our lives. The crowned cranes tell us that we are all one: they do so particularly when they take wing in order to guide us away from life's many dangers.

There's an open area beside a great generous expanse of river that's not actually a river at all. It's actually an immense winding lagoon, once part of the flowing Luangwa but now cut off by the inevitable forces that work constantly to change the place. It is a river flowing from nowhere to nowhere, and it's called Luangwa Wafwa, or the Dead Luangwa, grim name for such a lovely place. So sit here awhile and watch the crowded crowned cranes drifting away from you, as the zebras feed their plump selves, the stallions self-importantly herding the little breeding group together, little herds within a bigger herd, while the lithe lyrate impalas tread their delicate hooves on the bowling-green grazing lawn. Here the crowned cranes will slowly and inexorably vanish, muttering to each other in two syllables. If by any remote chance in this idyllic spot, you fail to notice that we are indeed all one, don't worry. The cranes will remind you.

70

Wet, Wetter, Wettest

Safety belt fastened. Legs astride the machine. Feet on the ridiculously fragile foot-rests. The propeller about six inches behind my arse, my pockets emptied of anything that might fall out and bust its delicate blades. Not scared me – oh no – veteran of half a dozen flights by now, spread out over the years. Little engine fizzing away with the pomp of the power-tool your neighbour insists on using on a Sunday afternoon. John Coppinger, the pilot, grasped the steering bar. "OK?" The voice reassuringly clear in the headset.

"OK!"

And, after a brief run-up, the micro-light took to the air and I was looking down on the Valley. Water, water everywhere and every drop to drink. Drink in the river, yes, obviously in the river, now as wide as the Thames at Westminster, over-spilling those sandy cliffs I had walked on, where I had sat to drink beer, where lions had lain to digest buffalo. And water everywhere else: the oxbows shining in sunlight, pools and puddles and ruts and streams and streamlets. The sand-rivers had become water-rivers, flowing with stern purpose into the mighty Luangwa. Pale gold had become blue-silver: the rivers were worthy of their name once again. There were temporary pools and puddles everywhere: I love the golfer's expression "casual water" – as if the water were leaning against a lamp post smoking a cigarette, or looking its semi-best, smart-casual in chinos and polo shirt – water that is relaxed, at ease, enjoying its moment. The parched, arid Valley had become waterworld: a soft, volatile place, changing shiftily even as you watched, the river carving out a new course, new lagoons being created. It was a land transformed.

And the waters were the waters of peace. At least they were for the herbivores. No longer did anyone need to tread the corridor of uncertainty – the daily walk of terror to the river – for now the pubs were open the length and breadth of the Valley and a 24-hour licence had been granted. You could more or less drink where you stood: no need to go in amongst the lions and play Luangwa roulette, the great game that someone always loses. The Valley was kind and gentle and forgiving: winter-time and the livin' is easy. This was March, the time of the water.

The tawny land, the lion-coloured land, had vanished, washed away in lavish tones of green and silver-blue. From the micro-light we looked down on fat elands, fat elephants, fat zebras, fat kudus, wishing perhaps there was a fat rhino or two to complete the picture. People used to say that the elephants left the Valley floor during the wet season, vanishing into the Mchindeni Hills, but from the air it's clear they do no such thing. They're just hard to see. Vehicles can't go where elephants go, especially not in the wet. There are all-weather roads in the Valley, gathered around the bridge mostly, and there are places where a good bush-driver can make a go of it, but there are vast areas where no one can take a vehicle in safety and plenty more – I've been to some of them – where travelling on foot is next to impossible. And there are places where you can only reach by flying. By being in a micro-light, say. Or better still, by being a crowned crane.

We overflew the Mutanda Plain, an area too squashy for any creature of any serious size. It is a place few human eyes have ever seen at this time of year, and it's where the crowned cranes breed. They gather in the heart of the wettest part of the Valley at this, the very wettest time of year, on land that can scarcely bear the weight of a mouse, where humans rightly fear to tread. There on the ground, in nests separated from each other as if someone had taken a gigantic pair of compasses to measure out each pair's permitted area, were the cranes, looking up at the buzzing intruders and shrugging an elegant shoulder, for John has got them accustomed and unworried bit by subtle bit. It is now, at the gentlest time of the year, that the birds of peace get on with the vital job of making more birds of peace.

71

Giving Blood

Ah yes, the perfection of nature. Can't go on about it enough. So let's consider another marvel: a fly so full of maternal love that she feeds her baby on milk. No, really, I promise you it's true. You don't have to make up wonders of the wild world: it's got far more than any of us could ever cope with; not weirder than we imagine but weirder than we are capable of imagining. It doesn't do the usual lay-eggs-and-buzz-off thing that most flies do: rather, it fertilises a single egg at a time; it allows the larva to develop within its uterus and it feeds the developing creature with a milky substance that it secretes from a modified gland within the uterus. You could say, then, that the fly knows both pregnancy and maternal care before giving birth to a maggot: one that very rapidly – within a few hours – becomes a pupa, from which it eventually emerges as a full adult. It's a fly that only feeds itself – in the usual sense of the term – as an adult. It is able to fuel so rich and demanding a life-cycle because of the equally rich nature of the food it consumes.

Blood.

Quite often mine. They bite with a sharp, painful snap. You don't itch afterwards, because they are not mosquitoes, which inject victims with an itch-making anti-coagulant as they take their blood-meal. It's over as soon as it's happened, but that doesn't make you philosophical about it.

They are tsetse flies: biological miracles that are capable of drinking their own weight in blood and then flying away. They subsist entirely on the blood of vertebrates, and as such, we humans are the equals of buffaloes and lions – all of us full of good blood – full of the old red red kroovy like it's put out by the same big firm, as Little Alex gloats in *A Clockwork Orange*. They are not as other flies: you can see this when they alight, and fold their wings, so neatly one behind the other, scissoring them shut with a fastidious air. Note the widely set apart eyes, too, and (obviously) the prominent proboscis. Iain MacDonald told me of a client he took on a 10-day walking safari in North Luangwa National Park, a man who was so concerned about being bitten by tsetses that he set off in full motor-cycle leathers. He was told (a) if they can bite through a buffalo's skin they'll have no problem with your leathers and (b) if you try and walk with this lot on you'll collapse of heat exhaustion on the first day. He did, too, and they had to help him back to camp.

They are innocents, these tsetses. They mean no harm. They only want a square meal and to be on their way, but they can leave behind a souvenir of their visit. They can, quite by accident, transport microscopic creatures called trypanosomes, and these can cause trypanosomiasis, a wasting disease, in their hosts. In humans, it's called sleeping sickness. In domestic cattle it causes reduced growth, lower milk production, weakness and usually death. And the Luangwa Valley is full of tsetses. They don't carry sleeping sickness, thank God, but it's impossible to keep cattle in the Valley.

And the result of that is that the cattle farmers were never able to take the place over. The tsetses kept it safe for the lions and the buffaloes and the antelopes and the hippos and yes, the rhinos too. So every time I get bitten by a tsetse, I ought to drop on my knees and give thanks to one of the most remarkable animals on earth. I don't of course. There's gratitude for you.

72

My Family

Neither Gerald Durrell nor my older boy Joseph could cope with school. Both were lucky enough to do without it – Gerry thanks to his raffish family, Joe thanks to his mother's organisational brilliance – and that, to most schoolboys, would seem paradise enough. But Gerry's paradise wasn't just the absence of teachers and pupils. When Joe, before his home education began, had a bad day at school, a not infrequent occurrence, there was always a way of ending the day on a good note. That's when Gerry came in to help. "Shall we do the birthday?"

"All right."

And so I would read the birthday chapter from *My Family and Other Animals*, in which Gerry persuades, cajoles, hectors and sweet-talks his family into giving him the birthday of his life, in which every single present was wonderfully relevant to the pursuit of natural history across the paradise island of Corfu, where the Durrells lived for five years before the second world war. Gerry gets everything he had campaigned, wheedled and plotted for, including a boat built by his brother Leslie, which he christens *Bootle-Bumtrinket*. This was followed by the best birthday party ever and, the following morning, Gerry takes out the *Bootle-Bumtrinket* for its maiden voyage with three dogs for company and an army of jam jars in which to collect the creatures of the sea.

The book is a story of paradise, and almost uniquely in books about paradise, the author has no interest in its loss. Durrell doesn't even mention the war. He says that the family move back to England for the sake of his education: "In desperation I argued against any such idea; I said I *liked* being half-educated; you were so much more *surprised* at everything…"

Thus in the concluding chapter, it seems that paradise is not so much lost, as temporarily mislaid; the book is not so much concluded, as put to sleep – gently euthanized as a good vet despatches a beloved family pet.

The book oscillates brilliantly from scenes of his loud bohemian family and the exotic glories of Corfu's wildlife. The chronology is superbly and unconventionally organised, so that five years fly by as one, and the literal truth is adroitly shifted about for literary purposes.

It's not, then, a story of loss. If you want a story of loss, read Douglas Botting's excellent *Gerald Durrell: the Authorised Biography*. It tells of a life that was often thrilling and frequently heroic. Durrell was a conservationist long before such ideas were part of the mainstream. His pioneering belief that zoos should play a part in conservation was ridiculed: it's now the orthodoxy. His lifelong campaign – unrelenting and unsparing – for wildlife was to lead him to breakdown and despair, followed by a personal redemption in his second marriage.

Wildlife conservation is also a story of paradise lost. Across history we humans have learnt to value the wild world at the rate at which we lose it. Loss is part of the human condition: we lose childhood, we lose innocence, we lose the sense of magical corners of the world, such as I experienced briefly at Edenever, and that Gerry experienced for five wonderful enviable bloody years.

But such childhood experiences are not always gone forever. They can turn up again. Sometimes you find paradise again as you find your lost wallet among the cushions of the sofa – but only after you've already looked everywhere else. And you wonder why common sense hadn't told you before that it would be there.

Every wildlife moment has about it a whiff of paradise, however faint. Every daily moment when the wild world intersects with the tame can feel like the re-finding of Alice's garden. It's a whiff of magic, and it's also a call to duty. As Durrell said in later life: "People think I'm trying to save nice fluffy animals, what I'm really trying to do is to stop the human race committing suicide." The means of suicide is not overdose or knife or rope or gun: it's more like blowing up the entire block of flats you happen to live in. Killing off the planet. Ecocide. An appreciation of the unlost nature of paradise is perhaps the only thing that might guide us away from that seemingly inevitable fate.

"'*Chairete*," he called in his deep voice, the beautiful Greek greeting, '*chairete, kyrioi*…be happy.'

"The goats poured among the olives, uttering stammering cries to each other, the leader's bell clonking rhythmically. The chaffinches tinkled excitedly. A robin puffed out his chest like a tangerine among the myrtles and gave a trickle of song. The island was drenched with dew, radiant with early morning sun, full of stirring life. Be happy. How could anyone be anything else in such a season?"

73

A Tribute in a Tributary

"I think we can get through," John Coppinger said. We had just returned to Tafika. After visiting the crowned cranes by micro-light we had had flown low over the river. I felt as if I could step neatly off the aircraft and simply skim along the surface of the water, so weightless did it all feel. We had turned off the main river along a tributary called the Chibembe and then, just above treetop height, we had cruised the length of a tributary to this tributary: the Chikoko, which is little more than a flood-plain channel linking the Chibembe with the Luangwa. The Chikoko was mostly hidden under a canopy of overarching trees, but peering through the gaps it looked as if we could get a boat through. "There's one tricky bit, but we can get as far as that and make a plan."

That, by the way, is an important African concept. Your plan is to make a plan. When you reach the tricky bit, you will work something out. The phrase conveys a notion about flexibility, resourcefulness, optimism, improvisation and an acceptance of the fact that life – certainly life in the bush – is seldom responsive to long-range assertions of the human will. "If we hit a problem, we've got Rod's Leatherman," I said.

Rod Tether, now an independent operator, was then working with John. He had recently acquired the multi-tool thingy called a Leatherman, and was carrying it in a sheath on his belt. This naturally required a fair amount of banter. "It's like a neck-tie," I said. "It serves no useful function other than to make it clear you're of the male gender. It's like the tail of a paradise flycatcher. It gets in the way, it's downright awkward, but you're not a true male without it." Rod, a fairly enormous fellow with shoulders that get stuck in doors – perhaps the reason he chose the outdoor life – took all this in good part.

And so that evening, we packed the RIB appropriately and set off. We were in no hurry. The engine muttered softly as we eased our way along the main river. The immense sandy beaches, on which so many dramas – a very few of them seen by human beings – had been played out, were now hidden under the vast restless sheet of water. At either side, and sometimes in mid-channel, giant trees had been ripped from the banks by the force of the water: a river I had seen when it no longer flowed and which I could wade without getting my knees wet. We turned off into the Chibembe. I had walked this route before. Walked, I mean, along the river, where I was now travelling by boat. It had been a winding soft sandy road, pockmarked with a thousand footprints, some ancient and indecipherable, others so sharp and freshly-minted that even I could read them plain.

We turned into the Chikoko, the engine's mutter now muted to a whisper, and our jolly company silent, the better to savour the wonders as the roof, the river's tent, closed in over our heads. Sweet Chikoko run softly till I end my song... ahead a monstrous tangerine cloud burst from between the columns of trees, a Pel's fishing owl, big as an eagle and improbably bright. And then the tricky bit. A tree across the water-course. It looked worse than it did from the air, because we could now see – and feel – the bits of it that were submerged. John nosed the boat into the branches and then we hauled away until the front end was halfway over. But below the surface, there was an obstruction we couldn't brush aside. Rod explored the river with his great hands: "It's not that thick."

"Well, you've got that thing on your belt…"

But Rod's mind wasn't on the teasing. His hands were underwater. There was a serious, rather awkward pause, in which no one seemed really sure what to do or say next. And then, quite extraordinarily, there was soft damp snap, as if someone had struck a lethal karate blow at the bad guy – and behold the miracle, we were drifting forward. The great Tether shoulders had twisted and knotted and then destroyed. The branch that held us was no more. There was a silence of

complete amazement. "You're a living…" John said, for once
lost for a word. "A living…"

"A living Leatherman," I completed. And then as we came
around the next bend another miracle: two bull hippos in full
combat and, even as we caught sight of them, they came
together nose-to-nose like two sumo wrestlers. And like
wrestlers, they came to a sudden climactic moment. One
hippo, his jaw seized and drastically twisted as his balance was
committed the other way, was flipped on to his back with a
thunderous splash – we saw all four stumpy tree trunk legs in
the air at the same time – and then both animals backed off,
deciding that the appearance of our boat signalled the end of
the bout. Beneath the canopy in this cool watery world, we
switched off the engine to enjoy the noisy silence of dusk in
the Valley. And a bottle of Mosi.

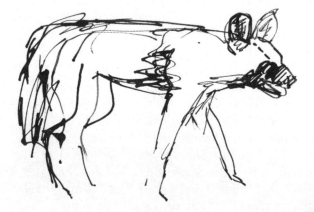

74

Fear in the Air

Take a walk in the English countryside. You'll feel all sorts of emotions and have a vast range of sensations, but you're highly unlikely to experience fear. If you do, the animal most likely to provoke it is a fellow-human. The nearest you get to a second dangerous species of mammal is a cow.

If your walk is otherwise supremely safe, you begin to think of nature as an inherently safe place. After all, we go to nature to get away from fearful things: the things that worry and oppress us like money and jobs and responsibilities. Here at least – in a wood, on a hill, on a coastal path – there's nothing to come and get you. No wonder nature seems like a bit of a paradise: a place of supreme peace where you are let off all the problems of the real world.

You look up and see the swallows crisscrossing overhead and think how joyful, how carefree their lives are: no mortgage, no job, no overdraft. But what they're actually doing up there is killing. Every sudden jink in their flight is a little death: an aerial insect is snagged in that sweet little beak and consumed on the go, the better to be ready for the next death. And while they do so, they are not feeling like swaggering conquistadors of the air. They're keeping a sharp look-out for anything that could kill them: most likely a hobby, a bird almost as agile in the air as they are, and with the edge on speed. Catching swallows out of the air seems like a pretty desperate way of making a living, but hobbies are small, neat falcons and they've mastered it. A hobby's entry into a feeding party of swallows tends to be pretty dramatic in either success or failure. They work best when descending from a dramatic height in the shape of an anchor or a Greek letter psi: striking before the swallows know anything's amiss. But swallows are wary birds and when they see a bird of prey, they shout out the alarm, a piercing double-note, and the bolder spirits will dive-bomb the intruder again and again until he has slunk off: the dashing hunter hunted by his own prey and haunted by worries of his own, for if he fails too many times he will die.

And never forget that swallows also face the most frightening journey. Those that hatched this year must – when aged little more than a few weeks – fly to southern Africa, and having done so, make the return to the cold sunlit north to breed. Most will fail to survive this double-trip. In the wet season you will see European swallows above the Luangwa and greet them as old friends, indifferent to the fact that they are probably Scandinavian or Russian birds rather than the ones that nest in my stables in Norfolk. But at least out in the Valley it's easier to grasp the fear that unites us all: all one!

So what about lions? What does a lion know of fear? A young male lion gets kicked out of the pride when he starts showing a sexual interest in the females around him, and then he must set out on his own, or sometimes with a brother or two. Away from the security of the pride, when the gorgeous, ferocious band of females feed and provide for all, the young males must find their own meat, often slumming it for small, pathetic bits of food like hares and guinea fowl. They must find a way to survive for half a dozen years until they are full-bodied and full-maned and ready to take over a pride for themselves. The smarter and luckier males will, of course, do their best to become sneaky fuckers in the course of their long, dangerous and fearful apprenticeship.

The prize, for the few that make it as a pride male, is immense: a great warm seething mass of luscious lionesses on which to sire the next generation, and no need to worry about anything but copulation, because these willing sexual partners will also kill anything that moves and the male, being more than half as big again as the biggest female, can help himself to what he likes. A male lion in his pomp is a poor hunter, being too big and too shaggy to be a great stalker, easily visible across an open plain. But why should he worry about that?

You'd think that he has nothing to fear, but that's not the case. The nomad males with their sprouting manes are ever on the lookout for a pride of their own. A pride male, however big, might be vulnerable to a joint attack. An ageing male will know that his days are numbered, that his grip is slipping. The day will come when he leaves the pride to slink off to die of his wounds. Then the incoming male or males will do as he did when he first took over: kill all the cubs so that the lactating mothers will come into oestrus at once and be ready to propagate his genes rather than those of his deposed rivals.

The wild world is a world of fear, just like the tame one. No one is immune. When Hathi the elephant told the story of how fear came in *The Jungle Book*, he was telling us that every paradise, every sacred combe that exists in the real world is filled with fear. It's a natural state of being. Fear is not to be seen as oppression or neuroses: it is simply awareness of danger. It is an ineluctable and irrefragable part of being alive and staying that way. We modern humans have less experience of danger than our ancestors, but when danger comes – and we survive it – we tend to feel that little bit more alive and, indeed, we sometimes find ourselves giggling like schoolgirls. Of course when we find the danger – well, we can make a plan.

75

The Emerald Season

It's a hard sell: come to the Luangwa Valley in the Rainy Season. Pick up the brochure on a wet day in January and you're going to say: oh yes, a holiday in the rain, that suits *me* all right, that's got me dipping into my pocket for a few grand. And it doesn't much help if you call it the Wet Season: it still sounds like an invitation to disaster; no, really, it's just a passing shower, let's stay in the caravan for one more game of dominos and it'll be a lovely day in a minute.

So tour operators have tried calling it the Green Season, which is fair enough, because it's certainly green enough, and it makes a nice contrast to the Brown Season – not that anybody calls it that because it's the Dry Season or, more often, just The Season. So they've taken it a bit further and tried to get people to call it the Emerald Season: well, who could resist that? It sounds as if it came straight from the Wizard of Oz, though your chances of finding a Cowardly Lion are remote. Norman Carr Safaris offer trips called Rivers and Rainbows: full of poetry, so long as you remember the fact that whenever you find a rainbow you are likely to find rain.

I was naively surprised when I learned that it doesn't rain all the time in the rainy season. It just rains more than it does in the dry season, which wouldn't be hard, since as you have seen, the dry season rainfall tends to be about zero – lucky to get nil, as the traditional sporting joke has it. Zambians dismiss much of the rain as "showers", though they look devastatingly heavy to European eyes. But "showers" don't last long: just a quick comprehensive soaking.

Oh, you say in your airy way, it's a hot place, so you just get wet. But when the rain comes, the temperature drops like an overturned hippo and it comes out of the cold tap, not the hot one. It gets right through to the skin and I've never found cold wet jocks agreeable. I take waterproofs and I use them, too. And sometimes it absolutely pelts down: the sky puts on a light show, fire against black and the rain falls not in drops or rods but in columns and the time it takes a waterproofless human to go from bone-dry to jock-dripping wet is about 1.5 seconds.

And it's wonderful, thrilling, glorious: gorgeous as a heap of hot lionesses is to a prowling sneaky fucker. It's a wonderful thing, a new way of thinking, a new way of looking at the world: rain is a joy-bringer.

How do you make life? Simple as $E=mc^2$. The equation is $s+w=L$: sun plus water equals Life. In our damp little island water is never too far away: it's sun we long for. We see sun as the great life-bringer: the sun that brings in spring, drives away the cold and fills the world with new life. But the life couldn't come if it hadn't rained all bloody winter, and if it didn't continue to chuck it down every now and then, even in the supposed good times, as we Brits have long complained about.

But out in the Valley, it's the rain that brings life. It's the sun that's always with you, the sun that drives people into misery and depression, the sun that makes sure October is called the suicide month. The rain, soft and gentle, sometimes hard and brutal, brings life cascading back into the Valley. Changing the colours, changing the values, changing the meaning. Perhaps the best name for the time of the rains is the Life Season. The problem is that this means the opposite time of year should be called the Death Season. The tour operators are unlikely to go for that, but you won't find a lion or a buffalo that disagrees.

76

Not-seeing Season

Here's another suggested name for the polarities of the Valley's seasons. We could call the dry time, the Seeing Season. That makes good sense: as the dry season advances, the vegetation dies back and you can see much greater distances. No one and nothing can hide. As we have seen, the casual water and the lagoons and the tributaries dry up and all the large mammals must drink from the main river, making for large gatherings of herbivores with their always faithful attendant carnivores. You can't help but see stuff: spearing fish in a barrel and all that. Those with cameras can go home feeling like the best wildlife photographer that ever clicked a shutter.

That makes the wet season the Not-seeing Season. The vegetation grows up thick and strong so you can't see the elephant that's 20 paces away and since there is water everywhere, the mammals can disperse and leave the dangerous river behind them. You can spend all day hard at it and count yourself lucky if you see the sort of thing you might find in half an hour in the peak viewing time. So I make a joke: it's great this time of year, none of those bloody animals to get in the way and spoil your day.

It's a bit like being in rainforest. I love being in rainforest because you never see anything. Well, that's an exaggeration, but you really don't see much. Life is teeming all around you, but it's all out of sight, hidden, generally, in the canopy. You listen to enigmatic whistles, sudden thunderous crashes, the whistling of huge wings. You hear sharp alarm calls and wonder what's been alarmed and who is doing the alarming.

And soon you get to understand that it's not the elusive jaguar or orang-utan or rhinoceros hornbill or harpy eagle that's the star here. It's the forest itself: the forest with you in it.

And in the Not-seeing Season you realise that the savanna is the star, the Valley itself, and that's when you begin to realise the true enchantment of the combe.

It's a bit of a connoisseur's thing. If you've spent a lot of hard-earned money and you're eager to see a lion before you die, you have a right to feel anxious until you've seen one. But when you've seen a thousand lions you are aware of greater things than the claims of a single species, however wonderful, however important. If you put the hours in, you begin to see something more than you ever thought you would.

And the Valley in the wet season is a bit like that. It's wonderful, but most people are only able to grasp the nature of its wonders if they have done all the drama and the lions and stuff already. If you have had the privilege of being able to put in the hours, as I have, you can begin to get a little closer to the true meaning of the Valley.

I was on a bank of a lagoon with Bruce Pearson, the wildlife artist. The sky was the colour of ink, but under these thunderous skies there was a strange glow about the vegetation. We were drinking Mosi beer, it being time for the sun to make its hurried descent into the trees. And a large lone bull elephant strolled out in his baggy-trousered ease, giving us a glance, canting his colossal head to do so, but without breaking stride; either at ease with us, or merely contemptuous. And he too took a drink, and as he did so he seemed to be the same colour as the sky, also lit with a strange glow. You don't see much in the Not-seeing Season, but what you do see, you treasure. I can see that elephant anytime I want to – and not just because it is recorded forever on the film at the back of my skull.

Bruce took on–the-spot notes and later made one of his
athletic watercolours of what was before us. It's on my
wall at home: it tends to be the first thing I see when I
wake up.

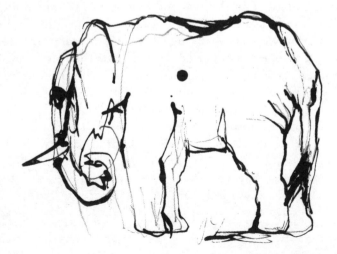

77

Caracal

Envy is a terrible thing, one of the seven deadly sins and not the most attractive. But what else was I supposed to feel? Though I felt a sort of joy at the same time. I was back home, a long way from the Valley, and talking to Gid from Kapani on the phone. She is an old friend, married to Norman Carr's son Adrian. She used to run Kapani lodge, where I have stayed many times, and now runs her own business out of Mufuwe. And she had just had an intimate encounter with a caracal. Had Mephistopheles been handy and in bargain-hunting mode, he could have had my soul for the price of that caracal.

Caracals are cats: not big enough to be big cats but far too big for your hearthrug. They're not terribly uncommon: they're officially listed as "Least Concern". But they're terribly elusive. I don't suppose there are that many in the Valley: they tend to be dry country specialists. They have adapted to get most of their fluids from the prey they catch. They are found all over Africa outside the Sahara and the rainforest, and into Asia as far as the top of India.

And I've never seen one. I've seen serval, the tall, slim, lithe cat of the grasslands, the one which, in classic bushman lore (otherwise known as bullshit), is able to walk through long grass without touching, still less moving a single blade. In fact, I must be one of the few people to have seen one and been disappointed: John Coppinger and I caught one in a spotlight and he confidently called it a caracal... until it emerged into plain sight and was covered in spots.

Cats are compelling. And there is a sense that in seeing caracal I would have reached a kind of completion or conclusion of my life in the Valley. Though, having said that, there are said to be cheetahs on the Chifungwe Plain, a remote area of the south park way off the usual tourist beat, but this has never been confirmed. It certainly looks like the sort of place cheetahs would like; I crossed the plain with Bob on one of our jaunts, but we didn't see much except grass.

Gid was glowing. I could tell that from her voice: one of those miraculous wildlife moments that makes you feels as if God has singled you from the crowd, on some divine caprice, to shower you with undeserved blessings. It wasn't one of those thrilling half-sightings, in which you can see but not see, in which you are half-elated and half-disappointed. This was the caracal in almost ostentatious mood: moving with long-legged grace, the glorious ear tufts plainly visible. They are wondrously athletic beasts; they can leap six foot in the air to bring down birds, it's a standard part of their repertoire.

I was filled with delight at news from the Valley and the sound of Gid's voice. I resolved to keep coming to the Valley – if I could possibly find a way of doing so – until I had seen a caracal for myself. Hoping that glorious event wasn't going to happen for years and years and years.

78

Whydah Still and Whydah

Visit the Valley in the Seeing Season and you won't see a bishop or a whydah – not unless you're a crash-hot birder, and even if you are you'll probably need a crash-hot birder as your guide (try Abraham Banda at Kapani). But make your visit in the Not-seeing Season and even if you're the worst birder in the world, you'll find them for yourself. The birds live in the Valley 12 months of the year: it's just that in the dry times they put on a cloak of invisibility. They are Little Brown Jobs or LBJs, as birders say, and it's hard to find a reason to bother with them when there are so many other marvels all around. But when the rains come they cast aside all modesty and become superstars.

The papyrus-beds turn a livid, almost unnatural green, a green so absurdly green you think it must have been filched from Augusta National golf course. And in the precise shade required to stand out from this violent colour you find – you can't avoid – the red bishop. Red: well, something of an understatement, that. It's red in the manner that a traffic light is red, red like the brake lights when the car in front of you makes an emergency stop at dusk. It flies like a flame across the eye-hurting wet green of the reeds: a flamboyant assertion that the rain has got his hat on and he's coming out to play. The LBJ has gone and in his place is a creature of utter flamboyance: coming out before the world as a heterosexual.

Then, as you move away from the papyrus into the open wooded savanna, you will find your whydah: for the male has put on a cloak of visibility. The paradise whydah has become as demented an attention-seeker as you will find anywhere in the

Valley, and you easily identify him because he takes about ten minutes to fly past. The male celebrates the coming of the rains by sprouting a train of feathers twice as long as himself, so that he flies with the air of a goods train: a really useful engine towing a couple of wagons behind him.

These two birds go from inconspicuous to glaringly obvious in, it seems, the time it takes for the sun to go behind a cloud. They are unmissable to you and me, and a magnet for such dashing bird-hunters as the falcons, goshawks and sparrowhawks that live in the Valley. You can't miss the bishop, and the whydah is self-handicapped by his endless train. Daring death is their sexual strategy, for only the best of all males could cope with the disadvantages they have heaped on themselves, disadvantages of outrageous conspicuousness and, in the case of the whydah, drastically reduced manoeuvrability. That's why the females – for ultimately it is the inconspicuous females that run the show, because they are the ones that make the choices – select the males for their showiness. And when the Tourism Season starts up again as the green begins to fade from the Valley, the whydahs shed their mad tails and the bishops fade away and become ordinary members of the clergy, as the colour fades from the final frame of *Butch Cassidy and the Sundance Kid*, leaving behind only a sepia memory of their outlaw glories.

79

The Eureka Bird

When I was able to pick out the song of the willow warbler, I entered another level of being. But it wasn't about becoming a better birder. It was more like a Zen experience of sudden enlightenment. I'd tell you what it was like, but you'd need to be enlightened yourself – and if you are, you don't need me to tell you. It was like one of those flashing non-verbal unthinking leaps of understanding that come with exposure to one of those Zen puzzles they call a koan.

A favourite story: a Zen monk is pursued by a tiger, so he leaps off the edge of cliff, grabbing hold of a stout plant as he jumps. He finds himself hanging above a thousand-foot drop with a tiger snarling down at him. He looks at the plant he is holding and from it he plucks and eats one of its fruit. And says: "What a delicious strawberry!"

I was enlightened by the song of the willow warbler: in an instant universe-crossing leap I was aware of the wild world as never before and of our own inextricable involvement with it. As the song invaded my ears and filled my mind, I was aware of the continuity of human and non-human life: a bird and me, linked for all time by the place we live in and the history we share, linked by ecology and evolution. No more dualities: no more dividing the world into human and non-human worlds.

I was led to this enlightenment by way of the Valley. All my time with Bob and with others had increased my awareness of absolutely everything – especially birdsong. It wasn't that I got better at spotting birds: I got better at understanding life. I could hear it going on, and because I could interpret the sounds, I could understand something of what they meant. And what they meant was that life was continuing and the earth was revolving through the day and orbiting around the year, and that I was a part of it. It was the willow warbler that brought me to this: not from the flamboyance of his song but from its simple understated nature, its subtlety, the fact that it can be so easily overlooked, the fact that, though I must have heard the song a thousand times, I never knew it for what it was. Now I did; this was not just a song, this was my *eureka* bird, the bird that gave me the key to the whole world.

I was reminded of one of the songs of the Incredible String Band, echoing back to me across the years from my hippy youth:

> *Always looking looking looking for a paradise island*
> *Help me find it everywhere*

And this cosmic revelation of the song of the willow warbler – this sudden and startling ability to pick this song out from the great cacophony of spring – changed me forever. It didn't make the entire world a sacred combe, or a paradise island, but it was a step in that direction. On Monken Hadley Common in Barnet I could hear this bird of paradise singing its sweet lisping song in late April, bringing not the rains but the sun and the spring in its train, and I was able to savour the truth and meaning the song brought with it. And like a sacred combe the song also had a dimension of secrecy: perhaps it was only me that knew the song for what it was. Secret and sacred are very close, not only in sound. But though I savoured the secrecy as a lover savours the secrecy of an illicit affair, I also felt a strong lust for the missionary position. I wanted to have this moment, this song, this bird, this paradise island, this universe to myself – but there was another part of me that wanted to share it: to allow others to enter in to the same sense of enlightened wonder. I've been writing about it ever since.

80

Listening to the Land

A landscape can be appreciated with your eyes closed, for it exists in sound as well as vision. Many of the most important parts of a landscape are invisible. They used to be beyond my reach. They were awaiting my moment of enlightenment, the moment when I was first able to go beyond the tyranny of sight. Naturally – observe the *mot juste* – I learnt how to do that in the Valley. You lie in your hut on the riverbank in the dark, surrounded by the tinkling of frogs and skirring of crickets, and the noisy silence of the bush rocks you to sleep. The lion's roar sometimes comes whispering from miles down the river and sometimes it seems to be right outside your door. Sometimes it is. The whoop of hyena sets the tone as nothing else can. Owls – and you soon learn to tell one species from another by the different hoots – love to be vocal, because they are creatures of the night and at night, hearing becomes more important than seeing.

Even in daylight, your ears tune in. It's hearing, knowledge and a bit of imagination, that enables you to see round corners and through impenetrable chunks of bush. You can hear an elephant snacking before you walk straight into him. It helps if you can recognise the calls of the oxpeckers feeding on exoparasites on large mammals. The sound tells you that a buffalo, maybe two, maybe, far worse, just the one, is on the other side of that clump of *Combretum* bushes. This is an ancient life-saving skill that our ancestors certainly possessed. Humans first began to walk upright on the savannas of Africa, and those that couldn't perform the oxpecker trick wouldn't survive to become ancestors. Knowing birdsong is a survival skill.

And here's a strange thing. Arrive in the Valley in the middle of the wet season and the place is unrecognizable. And it would be if you were blindfolded. You find a completely new soundscape, one dominated by woodland kingfishers: handsome birds that are highly territorial and musically somewhat unimaginative, with a loud, far-carrying descending trill, repeated again and again. Woodland kingfisher is not the contradiction it seems: plenty of kingfishers live away from the water and make swoops – not into the river for fish, but onto the forest floor for large insects and small vertebrates. The kookaburra is the most famous of this group, and the most famously vocal.

When the river pulls away from the banks to reveal those long sandy beaches, the waders move in and fill the hollow bed of the river with their calls, sounds many British birders associate with bleak estuaries on January mornings. But as the Luangwa returns and fills up the banks to the top and beyond, so the wading birds shift their ground, there being nowhere to wade.

Other sounds come to fill the silence: the homely twittering of European swallows can be heard over the Luangwa Bridge as they mix unconcernedly with wire-tailed swallows and lesser-striped swallows that stay here all year round.

It's not easy to impress a crash-hot Luangwa guide. I can do it, but only do it in the wet season, so it's worth going just for that. I have been going into the bush with Abraham for years now; Abe is the top guide in the Valley and lodge manager at Kapani. I have always relished his quiet ways, his profound knowledge, his skills at communication, his love and understanding of the Valley, and his sense of the responsibility all lodges and all guides should have – and also his great sense of humour and his companionship. I remember the first time we went into the bush together and as we walked, I heard a bird call I didn't quite get. I knew I knew it but I couldn't place it. Abe waited a courteous 15 seconds and then supplied the name. Say, kurrichane thrush. Abe really is as good a guide as has ever worked.

But there, in a scattering of thorn bushes that looked from a distance not unlike a patch of hawthorn scrub in England, came a voice, strong and marvellously sweet, assertive yet somehow shy, filling the air. It was a soft Valley morning. It had rained a little overnight, washing the dust from the air and filling the Valley with a subtle moistness that feels homely to an English person, but is most uncharacteristic of a valley where the sun is so ferocious and the heat is so dry that you never seem to sweat, it burns straight off you. But on this gentle day, the song was appropriate to cool, moist days across the world. And Abe, though he gives little away, was puzzled. This is not a bird you hear every day in the Valley, though he knew that he knew it all the same. There are times in England when I hear it all day, so it would be highly to my discredit if I wasn't able to supply a name at once. Abe was ever so slightly vexed with himself, because his speciality is omniscience. So I gave him a courteous, oh, maybe 20 seconds or more.

"Garden warbler."

A garden warbler is one of the loveliest, and one of the least showy singers, of an English May. They make their way far south for the winter, and will often lift their voices in song. How absurd it is to think of these tiny little scraps of feathers travelling 5,000 miles, if they are Scandinavian birds, arriving in this utterly unSwedish place, no Ikea for miles etc, and yet feeling as at home as you are in your own garden. And I, for once, had given Abe something wild.

81

The Golden Age

There are two standard locations in time for the Golden Age: the past and the future. No one has ever said: "The Golden Age is now: precisely what we're living in." It's always lost, perhaps irrecoverable, or it's ahead, just out of reach, perhaps unattainable.

I grew up in the 60s, a time when people, bizarrely, often presented WW2 as some kind of Golden Age: at least, one of inspiringly shared endeavour and bracingly shared hardship. Towards the end of that decade as I grew, to use the term loosely, up, the spirit of the time demanded a belief that everything was changing forever for the better: the Age of Aquarius was dawning and we would soon all be able to see the Krishna colours on the wall and find that paradise island everywhere.

I no longer strive towards such things – well, not in the same way – but the Golden Age left its mark on me. How else did I get to be a conservationist? Part of all conservation is an attempt to help the world, humans and all, return to a state in which biodiversity and bio-abundance are no longer under threat, so conservation is backward-looking. At the same time, conservation is an attempt to improve and enrich life in a manner that is sustainable over the long-term: so, in that way, conservation is essentially forward-looking. Conservation has to embrace this contradiction.

The map of the historical range of lions is a kind of cartographic tragedy. Three thousand years ago lions were found in Spain, France, Italy, Greece, the Balkans, Turkey, the Caucasus, the Middle East and the entire length of Africa, apart from the Sahara Desert, which was then much smaller. Now, if we want lions, we have to seek them – and we find them more or less confined to National Parks, where they survive because humans want them to survive.

There are wild notions about restoring wolves to Britain, perhaps bears – and what about lynxes? And always, a great enthusiasm for such notions is matched by atavistic opposition. The white-tailed eagle went extinct in Britain in the early 20th century. Captive-reared birds were released back into the wild at the end of that century. They are now established in parts of Scotland – and a fine sight they make, too. The first time I saw one, it was as if a giant had thrown a wardrobe off the mountain-top: a vast sky-blackening shape of uncompromising wildness. After this great success there was a proposal to release more eagles in East Anglia. It provoked a response close to hysteria. Farmers said the pig industry would become unviable, for the eagles would carry off their piglets. Fact: there are 400 pairs of white-tailed eagles in Germany and somehow the sausage industry is still functioning, but what's fact and logic got to do with anything? And of course, they won't just take pigs. They'll eat your pet cat. No dog will be safe. And how soon before they start to prey on toddlers? Fish and carrion are the eagle's main foods, but never mind.

There is a crucial difference between a Golden Age and a sacred combe: the sacred combe can exist in the present, in hard physical form. It doesn't have to be a place we once knew, or a place we hope to find some day, perhaps after death. There are times – and I believe we have all experienced them, that it is a common experience of humankind – when we have visited the sacred combe and known it for what it is. And because of that we can all live – at least for a few moments – in a Golden Age.

82

A Sacred Combe
in Redditch

I have lived through the Golden Age in Redditch. I have dallied in a sacred combe on the fringes of the Black Country. In the very middle of a classic '60s housing development set about by light industry and the white noise – actually more like black noise – of urgent traffic, I have been able to savour a whiff of paradise.

"I never knew it was there," said Stan. Stan is Rob's dad: Rob – Roob in the family – is my brother-in-law. Stan's in his 90s and has a next-door neighbour, John – busy, bearded, garrulous. And a man who got the sacred combe thing much as St Paul got Christianity.

The suburb of Winyates is one of those clusters of houses that seem a little cut off from normal life: a place dedicated to the preservation of the endangered species of suburbanites. It's green enough, in a lawns and lollipop-trees kind of way: a pleasant spot if you're not inclined to be snobby. John's house is a bit like 4 Privet Drive, Little Whinging, Surrey: home of Harry Potter's reluctant guardians. And yet a few paces away there is a place as filled with magic as the Forbidden Forest at Hogwarts.

A few years back John paid a visit and was struck by the thunderbolt of love. He discovered Ipsley Alders: an 18-hectare island of specialness totally surrounded by Redditch. The wild world lay just beyond his front door and now he was part of it. More or less at once he became a volunteer for Worcestershire Wildlife Trust and now counts it a day lost when he fails to take a stroll round the site. Keeping an eye, helping out, drinking it in.

For years John's been telling Stan to get me to pay a visit. Stan's been passing this on to Roob and Roob's been presenting the whole thing to me without prejudice. Almost with apology. I had managed to avoid this obligation for some years, not least because South Norfolk to Redditch by train is a wouldn't-start-from-here sort of journey. But Cindy, not for the first time in my life, put me right. She read an early draft of this book and said: "You're telling your readers that they can find a sacred combe if they're rich enough to go to Africa or lucky enough to live in Norfolk. That's not fair."

"But that's not what I mean to say."

"It's what you seem to be saying."

The realest, deepest and most important truth of this book is that a sacred combe is wherever you want to look for it. The need to find it comes from an impulse deep in humanity: an unstoppable urge to get closer to the non-human world. You can find a special and wondrous spot anywhere you're inclined to invest love, hope, need, desire, time. It's in your garden; it's in the park; it's on the railway embankment you pass on the way to work. So John took me to Ipsley Alders.

Ipsley Alders is mostly very wet underfoot and very peaty, fed by a spring of water the colour of whisky. It's pretty open, with rough grazing for cows who don't mind muddy hooves. There are a few ponds and an outer ring of trees. It's managed by the Worcestershire Wildlife Trust. Most of the year it's grazed by cattle. Michael, the volunteer warden, was there to show me round, along with Andy and Wendy from the Trust.

A buzzard flew over mobbed by a pair of crows. Chiffchaff hymned out their urgent message of the start of spring, for this was the first really warm day in April. A nuthatch gave a series of penetrating referee-whistles from the canopy. And across the reserve the acres of soft, wild ground were still, even after a dry spell, deeply hostile to human feet, infinitely capable of sucking the snuggest-fitting Wellington from an incautious foot. This was not a landscape tamed and civilised and made safe: here was atavistic nature in the heart of Little Whinging. "The other year we had three singing grasshopper warblers," Michael said with justifiable pride. I've been longing to find a gropper at home in Norfolk but it seems they prefer Redditch. Can't entirely blame them, because this is a wonderful spot.

It's wonderful even though it's right in the middle of a built-up area; it's wonderful precisely because it's right in the middle of a built-up area. It's an oasis for wild creatures and likewise for tired human souls crying out for a bit more wildness. We walked past a den made by local children: how many children get to make dens in the 21st century? A child who has never made a den has missed something of profound importance in life.

Of course there are problems with a wild place so close to so much humanity. Litter is a daily curse. Supermarket trolleys. A bird hide became a drinking den, so they locked it up. Then it got torched. Antisocial dog owners. Illegal fishermen. The odd rough sleeper. Andy found the abandoned hide-out of one such, and inside it, a revolver.

So it can be a trying business, running a wild place surrounded by so much tameness. Humans do tend to leave their mark. But all that makes the place more extraordinary. Every day is a kind of miracle: a glorious underdog victory for nature – that great old team fallen on hard times and now reduced to non-league status – against the rich and vulgar Premier League club of human development.

Blackcap were singing their sweet, rich song from a tiny fragment of ancient woodland, a thin strip of fine trees running along a manmade ditch almost as old. A tree-creeper called from overhead. And then, in a thin belt of conifers I performed a manoeuvre all birders are familiar with. I halted in mid-stride, mouth slightly ajar as if the better to drink in the sound, eyes raised uselessly at the impenetrable canopy, ears doing all the real work. And mind and memory too. Yes. Yes, listen. Listen hard.

"Firecrest!"

It's an enchanting little bird, wearing, like The Crazy World of Arthur Brown, a flaming headdress, and in season greeting the world with a thrillingly thin, high song. Michael, a deeply experienced ringer of birds shook his head regretfully. "Can't hear it." Many people lose the upper register of their hearing with age. I was the only one of the party that heard it. It was a new bird for the reserve, even if it's not going to make the official list just yet. "I'll stake my reputation on it," I said cheerfully. "Or we could raise the stakes to a quid, I suppose…"

It was also the first real butterfly day of the year; we saw brimstones – the classic butter-coloured butterflies – again and again. Later in the year it'll be full of dragonflies, with the emperor – a really impressive animal – the acknowledged star.

We worked our way back to the entrance and as we left I felt one of those small shivers, the kind that comes when you make too rapid a transition from one environment to another. A small hint of the bends. These urban and suburban reserves have something of the Tardis about them: bigger on the inside than on the outside, and playing odd games with your head in the matter of time. Had I gone backwards in time or forward? Did I spend a couple of hours in there or a couple of days? Or centuries?

I could still hear chiffchaffs and a nuthatch as we took to the cars and went to the pub for lunch. John with an I-told-you-so expression, one he was richly entitled to wear. He had shown me a special place, a sacred space, an enchanted combe: a place available to all, but one that very few people know about and make use of. Perhaps we are losing our ability to do so.

Yet the sacred combe is not an exclusive concept. It's as profoundly and recklessly democratic as sex or lunch or oxygen, and every bit as necessary. And not much harder to find: for there are miracles like Ipsley Alders everywhere you go.

It really was a firecrest. Trust me.

83

A Cathedral

We find it difficult in these secular days to connect with our sense of reverence. It is easier, for some reason, to grasp the opposite concept: blasphemy. We hesitate to call some favourite spot 'sacred', and yet we shudder at the idea of its desecration. The supermarket trollies of Ipsley Alders are a kind of blasphemy. Or say there was a place by the river where you once saw a kingfisher and every time you pass, you stop and hope to see another and even your disappointment is a beautiful experience – one that you might, in another age, call sacred. You wouldn't care to see the place scorched by fires and barbecues and littered with beer cans and crisp packets and condoms. You wouldn't want a caravan parked there, playing heavy metal at a million decibels.

The ebony glades of the Luangwa Valley inspire a feeling of reverence: certainly revulsion from the idea of blasphemy. Norman Carr is buried in one such glade. I saw my first leopard in the Robin Hood Glade, a few miles from Mchenja camp. And I have spent many an hour just sitting in the Mchenja glade at Mchenja camp, sitting still and quiet as life went on. On one occasion, a wart-hog and four clockwork babies passed in a line within ten feet of me and never even noticed I was there. On another, a bushbuck – a handsomely spotted male – strayed so close, I might have reached out to touch him. He noticed me and performed a double-take of epic proportions, letting out a vast bushbuck bark that shattered the peace of the glade, and set off as if a leopard was after him. Each ebony glade is a sacred combe within a sacred combe: a deeper and more powerful level of sacredness.

There was a time when Mchenja camp opened up for a few weeks in March, when the river was at its very highest. The camp had gone far beyond the stimulating basics of my early stays in the Valley: it was now a place of en suite showers and flushing lavs – no more of those thrilling midnight forays after an evening of Mosi drinking. At this time you could only travel to the camp by boat – the roads being impassable – and you gaze out, not at the vast sandy reaches, but at a fat, full river that flows with immense urgency. You don't want to paddle a canoe against it, that's for sure. The woodland kingfishers shout their descending trill from the trees behind camp, the river seeks to rip away more river bank from beneath your feet, and the lions roar petulantly up and down the banks, longing for the days when the livin's easy once again.

It was afternoon. Thin cloud, intermittent sun, pleasantly cool. And Aubrey – my old friend from that long stay – took the boat from Mchenja and chugged gently upstream. There was plenty of pleasant talk and banter as we took in the river and called out the birds. We passed a male puku with its antler comically covered in vegetation: it is the practice of territorial males to show off their ferocity and their mate-worthiness by horning the undergrowth and proudly wearing the debris afterwards.

And then all at once, with one accord, we all shut up. It was a truly disconcerting moment. For we turned off the main channel and we puttered – the engine very soft now, the progress very, very slow indeed – away from the river and into a wood. Now you're not supposed to be able to drive a boat into a wood. You may have heard that the best cure for seasickness is to sit under a tree; well, imagine how it is to sit in a boat weaving between tree trunks. I, who had been Robin Hood all those years ago, felt as if I was visiting a shrine to my own dreams.

Ebony trees are handsome things, with high, thick, bare trunks. It was like taking a boat through a cathedral, but then I wonder about cathedrals. Are those columns really there to support the roof? Or are cathedrals trying to look and feel as much as possible like a forest, like a sacred combe? The silence was a shared and powerful response to this impossible holy place. Ahead was a fish eagle. We passed a decent-sized croc. In one of the deeper pools, a hippo rested thoughtfully. In *The Magician's Nephew* there's a place called the Wood Between the Worlds: "Rich, rich as plum cake," one of the children says, when trying to describe it later. This was the same place: as rich, as quiet, as packed with meaning, as packed with life, as demanding of silence. Of reverence.

We emerged as people emerge from dreams, knowing that something of immense significance had taken place, but unable for the life of us to remember what. It wasn't a telling thing: my attempt to do so here feels ever so slightly sacrilegious. The better part of it I shared not with you, dear reader, despite all my best efforts. I shared it with the trees.

84

Vermin

We have little reverence for the commonplace. We save these swooping emotions for what is rare and special. Wonderment at the ordinary is a contradiction: one that can easily be set on its head by the use of drugs. Mescaline sent Aldous Huxley into raptures about the folds on his trousers; I stared in awestruck happiness at a hoverfly when lying in a hammock as I was coming down off my first acid-trip. The fact that I was (thank God) coming down added very greatly to my joy. It was one of those epiphanies: the everyday life of other species is far more wonderful than drugs – though it took drugs to give me that message. There is a sense in which finding the holy in the commonplace – in ordinary days, in the ordinary routines of family and working life – is what all religions seek to do. The greatest truths are not in the ecstatic experiences of wonder, but in the deeper experiences of the ordinary. That's largely what *Ulysses* is all about, which may be why it's been the biggest book of all for me, and one I found – and have found again on many occasions – singularly appropriate to the Valley. The Liffey and the Luangwa are the twin rivers of my soul: nothing less. Perhaps the biggest lesson I have learned on their banks has been about the blessedness of ordinary things. It's certainly the hardest.

So perhaps that's why white men in Africa attempted to shoot dogs: because, in the context of the wonders of the bush, they seemed offensively commonplace. Not feral dogs: African hunting dogs, sometimes called painted dogs. They're a different species from the domestic or feral dog, for that matter, a different genus: domestic dogs are *Canis familiaris* and African hunting dogs are *Lycaon pictus*. They're wonderful hunters, highly developed social life, and beautiful, in their odd way: every dog's fur pattern unique as a fingerprint.

They are probably the most successful mammalian hunters in terms of strike-rate: their percentage is up in the high nineties. This is because their cooperation is so good: they have evolved a tireless cross-country canter and they use this to the best possible advantage by taking turns to lead when they're involved in a chase – like athletes in a marathon forming an alliance to run down the leader – but they're also blessed with a sprint finish as powerful as that of Mo Farah. They have immense stamina, and thrive on relentlessness. They used to be found all over Africa, sometimes in packs of 100 at a time. I saw dogs find, kill and, in about a minute and half, devour a small antelope in South Africa. On another occasion, I spent an afternoon at a den – an occasion of exuberantly busy doggy body-language. They were both homely and wildly exotic. Eventually, a selected group set out on a hunt, leaving a few at the den to look after the pups. A great ritual of leave-taking swept through them, licking and sniffing, and then they set out with bouncing purpose. They looked as jolly as a bunch of dogs setting off for a trot round Streatham Common... and by the end of the afternoon, their walkies would almost certainly result in deathies for whoever they ran into. Then they would return to the den and the left-behind looker-afterers and the pups would all emerge together and solicit food from the almost-always-successful hunters, licking their faces and their jaws until the hunter obligingly served them with a warm, almost-fresh, meal.

The numbers of wild dogs have declined catastrophically. They need huge areas. Some National Parks in Africa are too small to accommodate the home range of a single dog-pack. They overspill the boundaries and run into populations of humans, and populations of the sort of dogs that humans tend to collect. Some wild dogs get killed by farmers, rightly fearful for their livestock, and others pick up diseases, rabies and distemper, from domestic dogs.

For years many people felt it their duty to shoot dogs on sight. Vermin, you see. Just bloody dogs: commonplace things, intruders in the economy of the bush. Call it gamekeeper mentality: the mad notion that nature can't survive without a bit of judicious management from a man with a gun. You'll have no songbirds unless I shoot the crows, the local gamekeeper told my neighbour, making you wonder how songbirds survived in the millions of years before the invention of the shotgun.

Dogs are now rare over most of Africa because they were once considered so commonplace that they needed killing. Thus they have become treasured creatures: very special indeed. And occasionally they are seen in the Valley, but not very often. When you hear the guides boasting to each other about dogs, you know they're talking about a beast you don't see every day.

85

The Curse of Specialness

For years I never realised how lovely a lapwing was and I'm sorry I ever found out. I wish I still thought they were pretty ordinary birds. Still, I suppose we'd better dwell on that loveliness for a moment: white face, thin whippy crest, great round wings that strobe black and white as they flip-flop across the sky. A flight of them passed over me the other day: a gentle black wave that switched direction and in an instant became a gentle white wave. A cry like an oboe that tells tales of moist places. In flocks they make Bridget-Riley eye-baffling art as they fly across the sky, each second a different and unique masterpiece. At breeding time, they perform aerial dances: rising sharply, descending, threatening to loop the loop, briefly turning into a giant butterfly and finishing off with a belly-scraping low-to-the-ground flight in which they wig-wag from side to side, black white black white black white, all the time accompanying themselves with mad sky-piercing oboe-shrieks, as if they were performing Stravinsky, as I suppose they are, *The Rite of Spring*, anyway. They were once common on farms across Britain: I remember seeing them in huge flocks as a boy and scarcely bothering to turn my head; it would have been as absurd as spending a day looking for sparrows. Just peewits.

But they have declined drastically, probably because of changes in agriculture since the early 1980s. So, for that matter, have sparrows. What was once ordinary has become special. I'm very much better at appreciating lapwings these days, but that's not entirely a good thing. They're kind of a favourite bird and I really wish they weren't. Or I wish I'd learned to appreciate their beauty by other means than their scarcity.

More and more creatures are becoming special. Inch by inch, step by step, familiar, semi-, or even wholly despised creatures become things to cherish. It's a pity we never cherished them when they were common. And among the things we cherish now, but never did before, is the countryside itself. We value wild places more than ever before, but that's because we've got so much fewer of them then we ever did before. The Romantic Movement had its roots in the Industrial Revolution. Before then, it seemed that we had a bottomless well of wild countryside and wild creatures, and so we behaved like millionaires: as if we'd always have the same ludicrous quantities of space and trees and wild creatures to do what we like with.

But now almost everything wild is special. The International Union for Conservation of Nature looks after the Red Data Books that track the status of individual species. It does so by a scale that goes like this: Least Concern, Near Threatened, Vulnerable, Endangered, Critically Endangered, Extinct in the Wild, Extinct. A vernacular scale would cover the same ground in a slightly different fashion: ordinary, special, gone. Blake said "everything that lives is holy". It's not yet true, thank God, that everything that lives is special – but there are more animals out there getting more and more special with every passing day and I hate it.

86

Last Footprint

And the Luangwa year turned, more or less literally beneath my feet. I was at Mchenja, and we went to the boat in the morning to find it lower than it had been when we tied it up the night before. It was like the first snow in the Arctic Circle, or the first drop of rain right here in the Valley: the still point of the turning year. The waters had started to retreat. The Valley was turning back into the place I knew a little better: the one with the narrow river and the wide river beaches, brown grass and carnivores rampant. A woodland kingfisher trilled out boldly, to explain that so far as he was concerned, the change was by no means complete.

It was a day of quite extraordinary privilege. I felt as if I was tiptoeing across the damp earth. I could feel the whole place held in the most delicate balance. There are some things that are only stable when they are in motion: dynamic stability, like a bicycle, or a racing kayak, hold it stationary and the whole thing collapses or turns turtle. This was a sense of stillness and movement at the same time, a perfect contradiction. A pause, perhaps, while the year cleared its throat. The stillness in the wind before the hurricane begins. The breath you take before making a speech. The pause before you place your arm about the shoulders of the girl sitting beside you. It certainly wasn't true that all things were possible: only one thing was possible, and that was the falling away of the waters and the rise of the dry season – but that's not how it felt. It felt as if the impossible was imminent. It felt, well a bit numinous, really.

We took to the river and the river retreated. When a stick or a branch extended from the water, it now bore a wet ring.

We made a trip into the Robin Hood Glade, but only so far. The boat needed to proceed on tiptoe, and far too soon, we had to turn back. There was a clear triple-note call; for a moment I was thrown, and then it came flooding back. Greenshank: one of the voices of the river beaches, silent, and for that matter absent, when the beaches are covered. But now it was back, knowing that the river would give it the gift of a beach – the first beach of the new season – in a day or two. The three notes were buoyant, confident, certain of its ground. The greenshank is a bird of the margins: between water and dry land, or with just its ankles in the water. It's a water bird that can only operate in the drier times of the year.

We passed a sandbank that showed maybe a square yard of sand above the rapid waters. It wasn't there yesterday. "Waiting for the first pratincole," I said to Chris. We travelled onwards; even now, it was time for Aubrey to proceed with far more caution than he had yesterday. The certainties had gone: this was now a river of shifting doubts.

We tied up and took a short walk. Got back into the boat again, and as I was doing so, I made a bad step. Chose a wrong bit of ground. Sank up to the ankle in soft riverbank clay; withdrew my foot with an effort and an audible suck. The pattern of the Timberland soles was clearly visible at the bottom of the hole. With a sudden dizzying lurch I realised that this footprint would remain until November when the rains came again. I had left a kind of fossil memory of my passing.

We returned to Mchenja. The sandbank was now big enough for 800 – I counted them – chestnut-winged pratincole, sitting in the intense proximity that brings them so much comfort and meaning.

There was another boat tied up at Mchenja. A visit from old friends: Abraham, Jess, Ade, Gid, Adrian. That night we had a kind of love-feast under night-sky. And next morning we set off back down-river, while the staff from Norman Carr Safaris began to pack up the camp. No one would reach this place by road for another two months; no one would reach it by river again for the best part of eleven months

The fish eagles sang out their eternal farewell.

87

A Twitcher at Large

Do you need an icky? A question incomprehensible to most of us. It's the way twitchers talk. Twitchers are a subset of birdwatcher just as the Amish are a sub-sect of Christianity and they share a tendency to extreme behaviour and hats. Twitchers are powerfully motivated by the lure of the list: by collection mania. And if a bird is not on your list, then you *need* it. You can have a list for the day, or the month, or the year, but for most twitchers, the list that really counts is the one for all the birds they have seen in Britain in the course of a lifetime. And if you need an icky, and you receive information – and these days there are a million ways of acquiring it – about just such a bird in a place accessible to you, then you go off and look for it. Your instructions will take you, perhaps, to a group of bushes on the North Norfolk coast, and there you will find a few like-minded individuals, recognisable from their telescopes and their hats. On an ordinary occasion, you will greet them with: "Much about?" But on this day, being dedicated to the pursuit of a single bird, you are more likely to ask: "When was the icky last seen?"

"It was right here half an hour ago."

So you wait and maybe you see it and tick it, maybe you see it as it flies away, and you know it was your icky but you didn't see it properly and you must, if you are true and honest to yourself, put this down as an NTV or non-tickable view, or maybe you never see it at all, which means you dipped out. And maybe you meet a friend from the same cult who was there half an hour earlier and got crippling views of the icterine warbler, so he had gripped you off.

And all that's fine and dandy, and as Miss Jean Brodie said, for those who like that sort of thing, that is the sort of thing they like, but it's never been my thing. I'm not snobbish about it, not unless I'm in the mood for teasing twitchers – any way you enjoy birds without killing them is fine by me. But I don't even keep a British list; for one thing it'd be embarrassingly short. These days you need 400 species on your list to be considered serious and, if you're that advanced, you'd consider an icky a tart's tick.

So picture me on a January morning in – count them – seven layers of clothing, walking down a path to a very specific spot on the Suffolk coast. I was going there because I had learned from the internet that there were three birds about. They had been there for two or three days and there was every chance they would still be there. And so I set off, not for a nice walk or for a good day's birding, to see what turned up: I set off to look for one species that had already been staked out for me. A twitch, in short.

I went to Trimley Marshes, a Suffolk Wildlife Trust reserve, and Mick the warden was there – we'd spoken on the phone the evening before. "They were showing well half an hour ago, but now they've retreated into the maize. Getting out of the wind. Can't say I blame them."

So I peered through his telescope – 'scope, to us twitchers – and I could see clear signs of movement behind the thickets of corn. That'd be an NTV, then. They were on the far side of the River Deben, a good quarter-mile off. So I did as twitchers do: I pretended I was actually looking for birds in general, and tracked gulls and crows and the occasional overflying mallard – constantly returning to the place where the target bird was Last Seen.

And then, as if it were the most natural thing in the world, three perfectly enormous and enormously perfect birds strolled out of the corn and looked straight across the water at me.

Cranes.

The same birds that had gone extinct in Britain centuries ago, the same birds I had seen in thousands in Sweden. Tall, and – even if in no mood for dancing, this being hardly the weather for it – effortlessly elegant.

You can never really quite predict anything in the wild world. It was in the 1970s that a small number of cranes turned up near Horsey in North Norfolk: quite literally out of the blue. It was as if a flight of angels had dropped in for tea. By 1982 there were five of them – and that's when, for the first time in 350 years, a pair of cranes bred successfully in Britain. They have since then spread out – in tiny numbers always – and now breed elsewhere in Norfolk, and they've bred in South Yorkshire. They've also been reintroduced to the Somerset Levels. Unlike their Swedish conspecifics, they showed no sign of migrating; they found they could make a living here all year round. I had visited the Norfolk Wildlife Trust reserve, Hickling Broad, on several occasions and seen the raptor roost with dozens of marsh harriers in the air at the same time and I had intimate moments with two species of owl. But the place is a favourite spot for cranes in winter and I had, well, dipped out.

Why did I twitch cranes? Why not any bird or every bird? Ah, but if you've seen cranes – any of the 15 species – then you know already. They would be wondrous things in any circumstances whatsoever, but here was a bird that had vanished and had then returned. It was as if humanity had not only been forgiven for centuries of sin: it was as if the harm our species had done was now in the process of being healed. It was like the return of King Arthur: as if the Golden Age had been restored. It was as if the enchantment had come back to the combe.

88

Combe Together

Fences are far more difficult when you jump them away from the herd. A horsemanly truth, and true in two senses that come down to the same thing. If you are riding a round in the showjumping ring, the harder fences are those in which you must travel away from the collecting ring – that is to say, away from all the other horses. A horse finds it easier to be brave and strong and purposeful when travelling back towards his own kind. An inexperienced or – in the jargon – ungenuine horse will feel a little sticky, a little reluctant, as you travel towards a challenging obstacle with the herd behind him; but he will feel much more eager – sometimes too eager – when asked to take on a fence that lies between himself and the others of his kind.

In other words, for a horse, an enchanted combe is not a fixed place but a mobile concept relating to the whereabouts of the rest of the herd. Home is where the herd is: a horse's meaning is in company. Horse-people will sometimes tell you: one horse is no horse. A horse has his meaning and his being in among others, and these others matter far more than any given spot.

It was this notion of defining yourself by togetherness that brought me back to Zambia again. I was in search of one of the greatest wonders of the planet. Bats! Ten million straw-coloured fruit bats come to Kasanka every year to feast on the fruit that is called into being by the rains. They roost in a few acres of forest, an upside-down city of chattering hard-flapping bats, seriously sizeable fellows with a three-foot wingspan, weighing a pound apiece and all stoned out of their heads on sugar and sociability. To stand beneath the sky-blackening spectacle of their departure at dusk, or their return in the pre-dawn, is like being at an aerial children's party, the air throbbing with their excited cries, a great rich urgent jelly-on-the-ceiling atmosphere.

And still they come and still they come, making great wheels, galaxies and nebulae across the sky, at evening heading out in endless thousands towards the great bonanza of fruit that may be 40 miles distant — and always there's one on the middle of the flock going back the other way with still greater urgency, having just remembered it left the gas on.

The effect on the viewer is deep and extraordinary: so many non-human lives, so many that branches frequently crack and break beneath the weight. Their chosen area of forest looks as if a hurricane had ravaged it. They are peripatetic beasts, these bats, following the rains up and down Africa, but never gathering in anything like the same numbers as they do in their annual visit to Kasanka.

I was there for three nights and on the last of them, beneath the great swirling cosmos of bats, I watched a martial eagle make a bull-charge into the midst while a group of a dozen tiny Amur falcons made fruitless efforts to grab a bat for themselves. I was then filled with a wild joy at the heart-piercing sight of so many millions and began to direct their movements like a mad conductor: now circle! Now return! Go left! Go right! Back a bit! You lot off to the fruit trees! Come in the next group! Standing beneath a miracle, alive with the joy of it all, filled with the same over-excited joy of the sugar-high bats above, I waved my arms to the skies and the bats and laughed and laughed at the wonder of it.

It is the largest bat roost in Africa and, in terms of biomass, in the entire world. For just a few weeks in the year, the fruit is available in colossal quantities and they come together in numbers borrowed from astrophysics. Together! Home is where there are many.

89

Cetti's

We were house-hunting. We were looking round a place and its garden in Norfolk when a Cetti's warbler sang out. "I'll take it," I said. Well, a slight exaggeration. But I knew things were more or less settled when I heard that bird. Cetti's – named for an Italian ornithologist so pronounced "chetty's"– are a little special. They aren't climate-change sceptics. They used to be rarities in Britain, real day-makers for list-building birders. But they responded to milder winters by expanding their range northwards and now they are found in wet places all over the bottom half of England. Their song is loud, designed to make birdwatchers jump. It's a powerful, deeply positive sound: a great defiant yell about the tenacity of life. You hardly ever see one: they like to lurk in a thick bush and shout defiance from its heart. It's been suggested that they are operating the Beau Geste Stratagem: trying to convince the world that there are far more of them than there really are – that way a visiting rival will decide that the odds are too great and fly on. Every bush seems full of Cetti's. They love noise, suddenness, secrecy. And as I considered this house and its surrounds, it seemed that the sound of the Cetti's was not just a message about Cetti's. It was deeper, richer than that. It told me that the whole place was wild, rich, life-laden. And I wanted to be part of it.

I also saw a marsh harrier flying over the marsh behind the house: as fine a bird of prey as you could wish to see. In 1971 these birds were down to a single pair that bred at Minsmere, the great RSPB reserve in Suffolk, but they are now an inevitable part of the watery landscapes of East Anglia: up to 400 pairs now breed annually. Their comeback was the result of the banning of certain pesticides, a lessening of persecution by gamekeepers, and sympathetic management of wetlands. So they too are birds of hope. And again, I wanted to be part of it.

A few months later I was sitting outside the house on a long, long June evening with a long, long whisky, looking out across the floodplain. We now owned the house and, thanks to the genius of my wife Cindy, a few acres of the marshland beyond it. It stretched, flat as if ironed, towards the river half a mile away. A Cetti's raised his voice with his usual impetuosity. And it was then I realised that the place reminded me of another flood-plain, another ironed-flat landscape, another place that was forever bouncing with life. Another place where I had drunk sundowners.

We manage the marsh for wildlife. Each spring we have heard eight species of warbler in song, and six of them breed there, either on our patch or just next-door. Sit by the dyke – I have a bench there – and most times, if you sit long enough, you will see a kingfisher, often whizzing fast over the water, catching the light and, it seems, burning a blue line across your retina in the manner of a sparkler.

It seems, then, that I have an enchanted combe of sorts just outside the back-door. Devonians wouldn't recognise it as such, but out in the middle of the Broads everything is either a marsh or a hill. The marsh before me is marsh all right; behind me the land rises a dizzying six feet or so to a meadow that's unequivocally hill. The land continues to rise maybe as much as 20 feet, and there, at the top, you find Hill Farm. A visiting friend criticised this for a half-measure: they should have called it Lookout Mountain.

So yes, it's a valley, and if it lacks anything as precipitous as the Mchinja escarpment or as demanding as the Mchindeni Hills, it still has a river at the bottom, higher land to either side and plenty of life. The other day a Chinese water deer – a deer whose barking is the most chilling sound in the valley, as good as a hyena, carrying a mile or more on one of those still, misty, spooky Broadland nights – strolled within a foot of my favourite sundowners spot.

I did a fairly unusual thing for me: I started to keep a bird list. All birds seen or heard from my place – garden, marsh or hill – could be counted. And so the list began to climb. Birds of prey: kestrel, sparrowhawk, marsh harrier, hobby, peregrine, hen harrier, red kite. Flocks of lapwings. Great gatherings of curlews flying over, calling so loudly and so sadly. A heronry in the wood just beyond our bit of marsh: you can hear the clattering and grunting and barking in the first few months of the year, and often see the great birds flying over on arched wings.

One bird remained: top of the wish-list. Possible, certainly possible. Though very, I thought, very unlikely. But always listen, always look. In a sacred combe anything can happen.

90

First Day of Spring

And London – all of London, every bit of it, especially the City and the West End – turns into a sacred combe. Or a bit like one, anyway. It's the first warm day of the year, the day where the sun is really pulling its weight for the first time since September. London responds as only a cold city can: with complete recklessness. Suddenly all the women are a single layer from nakedness – an expression of joy as much as desire – and men respond in the same sun-drunk spirit. Lofty hemlines, bare arms, missed-out buttons, ties discarded or worn at half-mast: it's as if the air itself were singing. The men, even the dullest, wear their jackets on their thumbs, and everybody's walk is different, shoulders thrown back, no more hunching against the chill, a bounce off the toes and a swinging of the arms. You almost feel that you could smile at a stranger without getting arrested. F. Scott Fitzgerald wrote of such a day in New York: "We drove over to Fifth Avenue, so warm and soft, almost pastoral, on the summer Sunday afternoon that I wouldn't have been surprised to see a great flock of white sheep turn the corner."

People bunk off work early and have a quick one at the pub that turns into three or four rather long ones, the crowds spilling across the pavements, pink arms, white shirtsleeves, a merriment in the air that's far more intoxicating than the bottles of Becks and the glasses of Sauv' blanc. These are not the hard-smoking, collar-turned up loiterers of the weeks before but a great mass of people brought together by the knowledge that spring is here, life can begin again and absolutely everything is possible.

I felt like this as I got up from my bed at Kapani in the Luangwa Valley. I felt like this precisely because the sun wasn't shining and because it was no longer pulling its weight. The previous day I had arrived, picked up at the airport by Innocent. The camp was almost empty of visitors: the first rains had fallen a week or so back and there were only two other guests. Many of the staff were on their break. I was there to visit old friends Gid, Adrian and Abraham, to look at a little wildlife and to breathe the good damp air of the Valley once again. Later that evening I was in the bar, quite alone, wondering where everybody had got to, when a voice called my name. "Who's there?"

"Innocent again."

"I wish to God I was," I said. But even as I said it quietly, softly enraptured with my own joke, I wondered: did I really wish to unlearn all the hard lessons that the years bring you? No matter: it was time to see Gid and Adrian and to drink Mosi.

The first rains had come perhaps a fortnight earlier. They had washed six months of dust from the air and replaced it with fizzing hope and inexpressible joy. The cuckoos were here: well, of course they were because they follow the rains and sometimes anticipate them. In the Valley they come in half a dozen species: the red–chested cuckoo was filling the trees with his triple note call: "Piet-my-vrou!" That's what the Afrikaners hear: Piet my wife! Three very distinct, loud, emphatic syllables, almost Cetti-like in their insistence on being heard. As I stepped from my cabin seeking tea – for it was mid-morning, I had needed a sleep-in after a long travel from the bat-colony the previous day – I found two species of butterfly, little shards of heaven, pursuing, as all butterflies do, the twin imperatives of drink and sex. Just like those crowds spilling out of the pub. I looked them up, and better still, managed to name them: redtip, and Zaddach mimic forester, black, white and crimson.

Innocent and I drove out to visit the mopane forest that lies on the same side of the Luangwa as the camp – so not in the National Park – and almost at once we encountered a small group of very muddy elephants, a breeding herd that came in all sizes, and they saw absolutely no reason at all to get out of our way. Nor, for that matter, did we. So we watched them a while: sharing the lovely morning with them, in mutual tolerance and appreciation. They were drinking from the many puddles that studded the ground – why? Because they could, I supposed. It was an active expression of their freedom from the river, their freedom from the daily walk of danger – danger at least for the little ones. Casual water – smart casual water, perhaps – was everywhere. It was as if it had rained gold: the most valuable stuff in the Valley was now free to everybody and in vast quantities – so much, you felt you'd never feel the want of it again.

That evening Abe joined me at supper. "Tomorrow and I think for the rest of the week, I will be able to take you out. Just the two of us." A good friend who also happens to be the best guide in Africa, and the whole Valley exploding with new life. That was all right by me.

91

The Great Comeback

And so, 15 years after the black rhino was declared extinct in Zambia, the rhinos returned. Zambia once had the third highest black rhino population in the world and lost them all. But in 2003 it had black rhinos again. Fenced in and protected, but at least back out there. The animals were translocated to North Luangwa National Park, which is less visited, and with a far lower human population than the South Park I know, so much better. The North Park is seriously wild: there are hardly any roads there and no permanent camps. And it's all unadulterated rhino habitat.

If your prime target in conservation is birds, you have a much easier time of it. You just get the habitat right and wait for them to fly in. So how long do you have to wait for a flight of rhinos?

The answer to that is five years. They came in not under their own power, but by Hercules aircraft. Five of them landed in North Luangwa National Park in 2003, five years after the declared national extinction, rather longer since the last rhino was killed for its horn in Zambia.

The North Luangwa Conservation Programme, NLCP, was up and running. It had taken four years of preparation to set it up, but now the great adventure was on: a glorious attempt to put the toothpaste back in the tube. It was established by Frankfurt Zoological Society, and it receives funding from organisations that include Save the Rhino, which is based in London.

Ten more rhinos came to North Luangwa in 2006, five more in 2008, and another five two years later. That meant that by 2011, there were 25 rhinos in a well-watched and well-guarded area of 220 square kilometres. They were kept in protected circumstances, initially in a *boma*, or timber-walled enclosure. Now the population is free-ranging, protected by the vigilance of people on the ground. Circumstances are not entirely natural, at least not yet, and the project will supply additional food to animals in sub-optimal condition and to lactating females.

It's not been straightforward. It never is. It's a process that makes considerable demands on the animals involved. They're not machines: the flight is difficult and the instant change of environment is traumatic. Such things can be hard for domestic animals, as horse-people will tell you. For wild animals, not selectively bred for docility, there will always be problems.

Some rhinos found it harder to make the change than others: after all, rhinos are only human. Some possibly contracted trypanosomiasis, others died in fights. And in the traumatic year of 2011, six of the rhinos died, more than 15 percent of the population.

"It was one of the hottest and driest years on record," said Claire Lewis, technical advisor for the NLCP. "All wild animals suffer under such extreme conditions. It was hugely disappointing, but we had to be pragmatic about it."

That is the problem with all small populations: there's no such thing as a small disaster. The project is still a fragile thing: it wouldn't take much to lose the lot, at the cost of a good deal of money and great deal more hope. A mother of a three month old calf was found dead, and the calf never seen again. Such things are hard for the people on the ground to deal with.

There was easy consolation to be found: six more calves were born. It's time to think about expanding.

And here's the miracle: so far not a single rhino has been lost to poachers. I spoke to Ed Sayers, the chief technical advisor of NLCP, and he said that, at the end of the dry season of 2014 there were 34 rhinos. That's a lot better than nought. Some of the fences had been taken down, the rhino area was getting bigger, and in five or six years, he was hoping for a population of 70-80 rhinos.

At this stage there would be a certain abnegation of control: the rhinos would be going beyond the ability of the project to give them constant guard on the money they have coming in. The outlying rhinos would become vulnerable; that's if you exclude the possibility of the world coming to its senses. So there are plans to move rhinos into other National Parks in Zambia: into South Luangwa, a few miles down the road, and to Kafue, on the other side of the country. "They have to start working on that right away," Ed said. "It took us four years to get ready for our first rhinos – and that was when the demand for rhino horn was a lot lower."

I've been in conservation for too long ever to think that a problem has been solved, that a species or a place has been "saved". But I still feel that defiant, almost treacherous glow of optimism when I hear of a good strong step in the right direction. It's awfully hard to put the toothpaste back into the tube, but someone's got to make the effort.

You don't do yourself or the world any good by pretending things are other than they are. No sacred combe is without its problems, apart from those that exist only in the human imagination. Gerard Manley Hopkins wrote of his desolation at the felling of a group of aspens: "After-comers cannot guess the beauty been." How I'd hate to address such words to the after-comers of the Luangwa Valley: how I'd like to pretend that my lovely valley was immune from the cares of the world, in the manner of Tom Bombadil's place – oh fuck, not another rhino. But I know it's nothing of the kind. The success of the tourism industry has been far more good than bad for that Valley, but all the same, it has caused an increase in the human population, and people fell trees to cook their food and so the buffer zone around the park is slowly disappearing: something that needs to be addressed sort of now. The resurgent demand for ivory has seen a desperate increase in the poaching of elephants. And of course, the rhinos are long gone from the South Park.

But how joyous it is to know that there are rhinos back in the Valley, even if they're in the more inaccessible part, and that there are teams of people who have made sure that for 11 years, these rhinos have hung out there and fed and fornicated and fought and died and given birth in the Luangwa Valley and that there are even possibilities of them moving south. Heady times. And if the continuing existence of unpoached rhinos in the Valley means that an uber-wealthy Vietnamese must endure his hangover without his billionaire placebo – well, I can find it within myself to cope with such a thought.

92

Bugles in the Sky

It never comes when you expect it, never when you're ready, never when you're in the right mood. I was in the middle of a series of hurried morning tasks, all those millions of chores that you have to do if you are mad enough to keep horses, when I saw a squadron of birds manoeuvring in a leisurely fashion in the big skies just behind the stable block. Five of them. Herons, I thought, though you don't often see them in the air in numbers like that: perhaps someone had disturbed the heronry that lies a couple of hundred yards away. Or perhaps the young birds had all fledged together and were trying a sort of massed break-out. But then, flying in tight unison, they wheeled through 90 degrees and moved towards the higher meadow, the one on the "hill". And I saw, with a great thump in the belly, that all these birds were flying with their necks extended.

You need to be a bit of a birder to feel that particular belly-thump. Herons fly with their necks tucked in, in the manner of a baritone sax; these had all stretched their necks straight out from their bodies. And that's not a subtle difference, not when you experience it: it's like a great braying trumpet call to alert you to the impossible. I suppose they could have been swans, but I knew they weren't: swans also stick their necks out and mute swans are always flying across our bit of marsh. But swans never fly in a leisurely manner: a mute swan is the second heaviest flying bird in the world and everything they do in the air must be direct and urgent and to the purpose or they'd come crashing out of the sky. These were no swans.

It was at this point that I actually heard a trumpet call. Well, more of a bugle-call, really, and I knew what that meant.

So I took a mad gamble and rushed back into the house – muddy boots all over the carpet – to grab the binoculars and, by a miracle, the birds were still in sight when I got back and I sighted on them and drank my fill. Long-necked, long-legs trailing behind, possessed by an airy elegance, and talking to each other in the language of bugles. Big grey birds against a big grey sky, birds of wild wet open spaces, birds that need a lot of wildness around in order to cope… and through the binoculars I could see with immense clarity that each bird wore a small, dashing red cap.

Cranes.

Well, of course they were bloody cranes, and I had them bang in the middle of the glasses for perhaps two minutes as they made their way, somewhat uncertainly, towards the marshes and fields they were looking to feed on.

A fine and fulfilling moment, but one not without a Proustian sense of disappointment. Proust was dismayed to find that the Venice he visited was not quite as Venetian as the Venice of his imaginings, of his hopes. And, absurdly, I was a little disappointed that my experience of the cranes was not still better, that my views were not of the kind to make non-birders gasp. This was magnificent enough, but magnificent only to a connoisseur. This vision would not convince a stranger that the ground here was sacred: it would only confirm this truth to someone who already knew. I tried to will them to stop, pause, to come down onto the common, better still to our bit, to feed and rest up, reconnoitre the place, consider it as a place suitable for the raising of the next generation of aerial buglers and dancers. But no, they looked down and decided to carry on, and though I was delighted by their visit, I felt the sting of rejection in the flyover as well.

Later that year I visited a nearby nature reserve, one too fragile to allow public access, and as I arrived I heard a deeper and more frenzied bugling than the calls cranes make in flight. And there, not a hundred yards away, was a crane with his beak pointing to the sky, hymning the air with his calls of joy and triumph. Here a pair of cranes were breeding for the first time in recorded ornithological history: for cranes, impossibly, in this ever-more overcrowded island and in this ever-more overcrowded century, are expanding their range and their numbers. And everywhere they touch becomes ever so slightly numinous.

93

The Fawn

The northern hemisphere spring comes slowly, gently, gradually, with many advances and almost as many retreats. But in the Valley the year turns on a sixpence: performs a showy handbrake turn and careers off in the exact opposite direction. The place changes as if it had just made a New Year's resolution; as if it had decided to be good forever after, giving up heat and broiling and clear blue skies and cruelty, instead being sweet and kind to everything that lacked pointed teeth or a hooked bill.

Abe and I drove across the Nsefu Sector and it was endless miles of celebration of water. Crowned cranes, of course, spoonbills, a gathering of 100 and more black-necked herons. In the early morning the elephants were bi-coloured: black below, pale grey above. The smaller, the blacker. They had crossed the Luangwa in the night to go raiding in the villages, for the wild mangos are in fruit, and they are an elephantine passion... an illustration of how difficult life is even here, when humans and wildlife find their interests clashing. Now they were back, recrossing the river with full bellies, the river water marking them again. Not a crossing they'd be able to make for much longer: the river would soon be impassable. A flock of green sandpipers, looking like sand martins with their white bums. A breeding herd of zebra with an ultra-charming leggy foal.

The agenda seemed to be set by the impalas. They synchronise their lives with the changing of the seasons, so that when the rains come in a great thunderclap and waterspout, it is as if it had rained impalas. Lambs, you're supposed to say, but they're nothing like lambs: they're surely the fawniest fawns that were ever dropped, with four legs like strands of uncooked spaghetti, absurd little hooves and that just-brushed chestnut golden coat. Every so often you find one still wobbling, not yet with the few hours it needs to be wholly comfortable on those brand new little legs.

That evening Abe and I climbed the Chichele Hill, and from its summit we could see the Mchindeni Hills on one side and the Muchinga Escarpment on the other: the whole Valley from east to west. The rains had washed the air wondrously clear and we could see more or less forever. On the plain spreading out below us like all the kingdoms of the world, we could see busy herds of impalas, pukus, warthogs, and among them cattle egrets and sacred ibis. Open-billed storks were digging for snails in the soft mud. This fierce land had given itself over to gentleness. Not that the snails would agree: they had lain buried and dormant beneath the concrete earth for months until the rain softened the world and freed them – only to make a snack for a bird that looks like a pterodactyl.

I felt that nothing could go wrong and I could do nothing wrong. There was magic in the air all right. I have seen so many wonders in this place, but nothing compares to an impala fawn, I thought, as I look down from this eminence at the tiny blobs of life below. Sometimes they stood looking as silly as a child, or an acid-tripper, looking out in solemn wonder at the falling of a leaf, the passage of a butterfly, the drifting patterns of the horizon. Sometimes they turned to their mothers with a frenzied head-down charge, their desperation for a drink so violent they seemed about to knock her off her slim and pointy legs and toss her onto the floor as a bull tosses a matador. And sometimes, for no reason that seemed apparent, they would hop and jump a few paces on unpractised legs; trying to see if they could count up to four and wondering if that's what impalas are supposed to do.

I felt that life would be complete if only I could pack half a dozen of them into my bag and release them in the garden at home, there to frolic and bounce and dance in a place where no harm could befall them. They look up at your vehicle when you stop, as if the skull were at least two sizes too small for the eyes they had been given, gazing at you in wonder as if you were God: so that all you want to do is stroke their sharp little heads and tell them that everything will be fine forever after.

In *Alice Through the Looking-Glass* there is a sequence in which Alice enters a wood in which she can no longer remember her name. "Just then a Fawn came wandering by: it looked at Alice with its large gentle eyes, but didn't seem at all frightened... So they walked on together through the wood, Alice with her arms clasped lovingly round the soft neck of the Fawn." This gloriously unexpected image of innocence, so lovingly drawn by Tenniel, is one that has always haunted me: another kind of Eden, another of those secret places in which you can escape not only the world but the human condition. Neither Alice nor the Fawn know who they are, so they walk in deep affection through the wood: "Till they came out into another open field, and here the Fawn gave a sudden bound into the air, and shook itself free from Alice's arms. 'I'm a Fawn!' it cried out in a voice of delight, 'and, dear me! You're a human child!' A sudden look of alarm came into its beautiful brown eyes, and in another moment it had darted away at full speed."

94

Doggedness

Abe parked on the Luangwa Bridge. It had been a good morning, long and rewarding and full of elephants. Lunch was beginning to feel like a good option. Abe strolled over to a vehicle from another lodge, also parked on the bridge, and talked to the driver. This pooling of information is an important part of professional life and it is also, for most of them, a passionate engagement with the wildlife of the Valley. Talking shop is one of life's significant pleasures: the better the shop, the better the talk. So instead of talking about what the sales manager got up to on the last conference and why business is down 0.5 percent over the last fiscal quarter, the guides of the Luangwa Valley talk about where the alpha male of the Kakuli pride has been hanging out and whether the female leopard from the ebony glade near Mfuwe Lagoon has been seen with cubs yet.

Abe walked back and climbed aboard. The sun was pretty warm, if a little off its peak of a fortnight back. Time to seek shade and rest and food and drink. But I could tell by the way Abe placed his hands on the wheel that something was up. He sighed, again an unusual thing for him. He didn't start the Land-Cruiser. I began to smile, very faintly indeed. Maybe the smallest smidgen of an adventure lay before lunch. I stayed silent so as not to break the spell. Eventually Abe spoke.

"Do you know we have missed dogs by an inch?" In his voice a mixture of personal and professional disappointment. He wanted to see dogs for the sake of the dogs. And as the Valley's top guide, he was mortified to have missed them.

I knew now what we were going to do. But Abe is never a man to be rushed. "Are you hungry?" Very courteous, Abe.

"Not especially."

Another sigh. "Do you think we should go back and look for them?"

"I do."

"So do I."

So we drove back to the Lupunga Spur, hurrying past all those lovely elephants and giraffes in whose presence we had lingered, past the place where I had managed to identify a small spotted sailor and the spectacular emerald swordtail – butterflies both – and past the place where we had heard two male emerald cuckoos taking part in a ferocious song duel, rivalling each other for volume and frequency and intensity until one, out-sung, backed down and flew away distraught. Pretty Georgie! That's what these birds are supposed to say: four great notes that seem almost to burst the singing bird into bits.

And then we found our dogs. For the first time, after God knows how many visits to the Luangwa Valley, I was looking at Zambian wild dogs. There were six of them, piled together higgledy-piggledy under a bush and panting in the heat. They fidgeted about at our approach, and shifted their positions, but they were content enough. They're not as accustomed to vehicles as the more sedentary predators like lions, hyenas and leopards, but they were able to get their minds around the problem. I was able to gaze on those radar-tracker ears, and admire the fingerprint uniqueness of each coat, and their uncompromisingly doggy ways. I remembered a man I had met in Zimbabwe years before, who had adopted a wild dog puppy as a pet, when he was boy. He told me how the dog became the perfect companion: the sociable instinct of a dog dominant even in an animal only distantly related to the dogs we humans choose to live with.

There is an affinity between humans and dogs, however wild. Dogs are not creatures of mystery; sure, these wild dogs are rare and elusive, deeply unpredictable and cover a massive home range – but when you come among them there is always something of your own hearthrug. Vermin no longer. We are able to enjoy their presence with joy; cherish them for their scarcity, their dogginess and for their sheer damned doggedness in hanging on in difficult circumstances and small numbers.

So we drove back to the lodge, me with a feeling that I had come a step closer to completion. I had seen dogs, I had hung out with dogs in the Valley. Another species to move off the wish list. There is a deep pleasure in that – and one flavoured with a distinct sadness. I want to see everything, but I don't ever want my relationship with the Valley to achieve completeness. The sense of mysteries still undisclosed: it's something that matters in every relationship, in the greatest of all books, and right here in the sacred combe. As St Augustine prayed for chastity, so I pray for completion of my understanding of the mysteries of the Valley.

Not yet, oh Lord. Not yet.

95

The Old Hippy

Everyone is familiar with the way that a word or a name or a notion will stalk you: appearing unbidden, vanishing and then turning up again. Statisticians understand this process is one of mere inevitability but, for the rest of us, such coincidences seem nothing less than magic in action. I came across this idea for the first time when reading something a bit Celtic and spiritual and highfalutin. It came, I confess, in something I was reading about the Incredible String Band, the greatest hippy band of my youth, the one who had sung about the paradise island and its potential ubiquity. These days the String Band counts – well, not exactly as a guilty pleasure but certainly one that requires a bit of defiance. So anyway, I made a mark, or mental note, because it seemed to fit the book I was writing. I meant to transcribe it into my notebook, but somehow that never happened. I expect that was because I resisted it a little. Too airy-fairy. A bit too acid-folk. The old hippy in me must be kept on a leash.

A few days later I came on the idea again, and by a most unexpected agent. My old friend Matthew Engel, former editor of *Wisden Cricketer's Almanac* and a companion in press-boxes across the world, had just published his superb and massive *Engel's England*, in which he visits every county in search of its meaning. Its soul, if you prefer, but Matthew doesn't go in for fancy stuff like that. His usual demeanour is of good-humoured but weary cynicism. He believes that if you expect the worst of people you are seldom surprised and never disappointed. His prose style is mostly built to match: understated, subtle, witty, deceptively straightforward. And if anything, rather lowfalutin. But this strange quest seems to have revealed – or perhaps released – a long-submerged spiritual side, and it's all the more vivid when coming from such a man, from such a writer. In the course of his Huntingdonshire chapter, he naturally paid a visit to Little Gidding. This was, of course, a TS Eliot pilgrimage, the name of the village being the name of the last of the *Four Quartets*. In the churchyard Matthew talks to a volunteer gardener: "When I work here in the winter I can see what Eliot meant. It's as though heaven and earth are very close together. There's a phrase that crops up in Celtic legends to describe somewhere like this. It's called a thin place."

So here's an alternative title for this book: *Ballad of a Thin Place*.

96

A Banquet

So naturally that afternoon Abe and I went straight back to the Lupunga Spur where the dogs had been panting hard in the heat – and in the cool of the late afternoon we drove straight into a drama of the highest order that the Valley can offer. Even as we turned onto the broad, flat outlook of the open plain we could see that the dogs were already up and at it, moving from that relentless ground-eating lope into a full dog-gallop. And yes, the target was one of those heavenly, gorgeous, innocent, just-hatched impala fawns, those fawns of vast innocent eyes and tiny wobbly legs, the loveliest and sweetest things in the Valley or Africa or the world. It was a mismatch of horrendous proportions: Real Madrid against the Rose and Crown; Usain Bolt against me; humanity against the rainforest. The contest was too brief to be brutal: one moment the fawn was a merrying child of gambols and promise and future, the next minute it was a meal.

Though that made it a thing of joy for the dogs. And then I realised that this was not the whole story: they had knocked down another fawn moments earlier, another earlier still. It was like *Blue Peter* for wild dog puppies: here's one I killed earlier. They made three little groups across the plain. There was no desperation in their feeding, everything had been far too easy for that. They commuted from one little corpse to another, graceful tails waving high in delight, and they shared and then they picked up the meal and played tug of war without urgency or rancour. The air was filled with altogether unexpected twittering cries, bird-like, the sounds of peak social activity among the pack.

As the hunger was slaked, the filling of bodily needs became less urgent and was replaced by social needs: for a dog, as for a human, a plentiful meal is an occasion for shared jollity, for strengthening bonds, for sharing the feeling that God's in his heaven and all's right with the world. Tough luck on the impalas, of course, same as it was tough luck on the snails that the open-billed storks were feasting on – same as it's tough luck on the turkey at Christmas dinner – but it was joy unconfined for the diners. Each dog went all around the pack, greeting each fellow-member with doggy sniffs and great sloshing face-licking, to clean off the blood and gore, for the pleasure of the taste, as a service to a pack-member, in the expectations of future favours, as a puppyish tribute.

Two of the dogs went in for some heavy scent-marking: lavish douches of urine splashing about the plain, one dog's offering topped effortlessly by another's. Then these two went to lie down together in the shade, on their breast-bones, heads up, watching the fun. I guessed these were the alpha male and female, big round ears making that absolutely distinctive silhouette. The other four, probably all of them younger, got on with a more active socialising: romping and play-fighting together in high content. It wasn't as if the hunt had taken much out of them, after all. They went through a series of classic manoeuvres: standing on their hind legs, locked chest-to-chest at an angle of 45 degrees, like Greco-Roman wrestlers at the Olympic Games, till this A-frame structure collapsed under the pressure and they could slink and curvet around each other taking mock-menacing bites at each other's flanks.

The sun was making its hurried descent and eventually the dogs, daylight-lovers, grew calm. Shockingly, or not shockingly, there were impalas and their fawns grazing and browsing a hundred yards away: one of the younger dogs watched with fascination, as if at a good programme on the telly. Then another dog stood, intrigued by this open display of interest. The first dog made a half-hearted trot towards the impalas. The impalas withdrew, briskly but without panic, playing the percentages, knowing that the dogs were not altogether serious, each one knowing that in a herd the odds are always against it being me that gets got, and besides, certain that if anyone is destined to play a part on the menu that night, it was not going to be an adult.

And so the pack of dogs decided to use the last of the light to move on and find a comfortable place for the night's rest. They gathered themselves up and trotted away, panting lightly, each tongue a-flap, teeth bared in a lopsided grin, just as a Labrador pants in self-delight after fetching a stick.

97

Being Quite Wise

The alarm on my phone is set to the gentlest and most natural sound it can offer: the chirping of crickets. It still sounded like a nuclear explosion when it started chirping at quarter to four in the morning. I could hear the voice of my inner Sergeant Wilson: "Do you think that's quite wise?"

But I'm an old hand at this. I refused to listen to the sergeant, got up with only a groan or two and dressed. I went downstairs and made a flask of tea. Rooibos tea, comes from South Africa, part of that unique fynbos flora, a taste I'd picked up in the valley, where else? I put on some outer garments: I'm an old hand at that as well, so I dressed in full winter gear. It was May – May in England, May in Norfolk, May in my own house – but no need to get carried away.

I'd timed it dead right. It was still night: just a fraction paler in the east, as it should be. I had hardly set foot out of the house when a Cetti's warbler crashed out its brief, sudden song. For a moment I thought about going straight back in, because there was a kind of perfection in this encounter. But I pushed on, across the marsh till I came to the bench by the dyke, and there I sat. The stiller you are, the more layers of clothing you need, and I intended to be pretty damn still. It was up to the rest of the world to do the moving: slowly coming awake all around me in its own good time.

There are two kinds of birders: May people and October people. I'm May. October people are drawn to rarities, to hunting, to accumulation, to listing, to sighting birds out of their normal context, to the development of first class observation skills and with them, the skills of finding birds in unlikely places, the skills of the bird-hunter. My friend Carl is like that. He's taken me on thrilling October forays when migrants fall like rain onto the bushes of the North Norfolk coast: canopy-dwelling birds like redstarts are flocking round your ankles; wrynecks sit quizzically in the open and icterine warblers – should you need one – lurk deep in cover, making enigmatic calls.

But May is my heart-time. I prefer to seek not the unexpected but the expected; the spectacle, the richness of this great season of life when every tree and bush and hedge and reedbed is in song. Even in the darkness the Cetti's sang his guts out and as the light began to advance, so the rest began to follow.

Cuckoo! The bird that was once the harbinger of spring is now more like a harbinger of doom. We hear them less and less. I've been through seasons without hearing a single one. And yet this damp acre or so is the breath of life to them: for six weeks, with May in the middle, the air is filled with their calls, and they punctuate family life. It's as if this were a place where no harm could come, where the bad things that happen on the rest of the Earth are forced to stop at the boundaries, like Tom Bombadil's acre; as if time itself has been halted at the boundary and the rare, harassed and hard-pressed cuckoos were back in noisy profusion; as if the last half-century had never touched the place and peppery old colonels were reaching for their quill-pens to write to *The Times* about the bird that had completed this cliché, this ancient English idyll. You could almost imagine one of the savage beasts of Saki's stories bounding in to set complacency about the wild world into panic-stricken flight – perhaps the hyena that the Baroness called Esmé. Cuckoo! The call of the male is simple and strident, being designed to carry vast distances. Unlike the sedge warbler now singing in the reeds just behind me, the cuckoo holds no territory. Rather, he flies from place to place in a likely area, telling all the world he is there and amorous and would deeply like to meet a female in a similar mood. He is seeking a fleeting relationship: love on the fly. Afterwards the female develops an egg inside her and looks for some suitably helpful little bird to take on the job of incubating it and rearing the murderous youngster that will emerge: usually dunnock, meadow pipit, pied wagtail or reed warbler.

In the wood a hundred yards away I could hear the bark and growl and clatter from the heronry: four nests there, I had counted on a visit to my neighbour's land. A tawny owl called out, as if protesting at the end of a too-short night. I poured rooibos tea, drank, observed a kingfisher fizzing along the dyke. This stretch of water is six feet wide, with a gentle flow, sheltered, quite deep – and very rich in life.

There are joys of the whooping, air-punching kind; this was a joy of the quiet kind. Even a smile would have been too demonstrative, too noisy, breaking the flow. Let the world wake up as it will, let the wild world sing out according to its mood; my job was to sit and drink tea and listen. Look too, as the light came flowing in as slowly and gracefully as a tai-chi adept. A barn owl, wonderfully spooky in the half-light, flew along the hedge-line of a distant field – a very gratifying sight because they got a hammering in the hard winter 15 months back.

When do you call it a day? A question that vexed me – but gently – in two senses. When should I get up and leave? And when does the night I got up on become the day I go to bed on? It's like surfing: always you have to ride one more wave. So I sat on for one more bird, one more small piece of wild magic. That's why you need to wear so many layers: to protect yourself from your own folly. So I sat on, hunched, chilled, entranced. Waiting for one more good thing. Just behind me a willow warbler raised his voice. I sipped more tea, listening to my *eureka* bird. I still didn't want to move. It was as if I was obeying an order to stay.

There was a hearty splashing up the dyke, out of sight. Could be a moorhen, or a mallard, but I didn't think so. It was the wrong kind of splashing. I felt a knowing smile seeking expression on my face, but I didn't move a muscle of face or anything else. It wasn't the moment for muscle-moving. But give it time, give it time. There's a slight kink in the dyke, which is guarded by a willow that leans out a little, so you can't see too far. You have to wait and see, wait and listen too.

227

And I'm pretty good at both these things. Then round the kink it came: vigorously chucking water about as if it had no need whatsoever for anything as wimpish as discretion.

A head burst from the surface about 20 feet away, and the otter's whiskered face was revealed, all its mind on fish and not a care for anyone who might be watching and drinking tea. Down it went again in one of those flowing, curving surface-dives, the characteristic sigmoid curve of the otter in full hunt, elegant bum briefly sky-pointing, a flick of the tail and then gone, but you can follow its course by watching the progressive disturbance of the water; up it came again, so near I could have reached out to stroke the pointy head with the boy-racer's slicked-back hair, and then once again it made that flowing S-shape and vanished, because there was a fish not far away and the name of that fish was breakfast.

The ripples died down, picked out in a silver now touched by the gold of the sun, argent tinged or, and I could hear the sound of the enthusiastic pursuit hurrying along the dyke and then growing silent. Breakfast-time. Yes indeed. The day had come. Life was everywhere. I felt quite wise.

98

Phoenix on Speed

A grim dystopic city of the future – maybe as much as five years ahead – a city that's mean and hard and ugly, and where the people are the same. "And I was mean and hard and ugly too." The narrator is a young girl who lives by stealing. One night she meets an old lady with a big fat bag and steals it – only to find the bag is full of acorns. And so she starts to plant. And plant and plant and plant. The book is *The Promise* by Nicola Davies, and I have read it to my younger son so many times that I know it by heart. "Green spread through the city, breathing to the sky, drawing down the rain like a blessing."

I met Nick when we both went to Borneo with the World Land Trust on a rainforest project, and it was one of those instant friendships: exactly the sort of thing that happens when you are a teenager, the sort of thing you thought you were too old for.

Biophilia. The human connection with non-human life. Ever patted a dog? Smelt a rose? Then you know what I'm talking about. I feel it in rainforest perhaps even more strongly than when I'm in the Valley, and I think that's because of the essential mysteriousness of rainforest. You don't see much but you know that life is all around because you can feel it. So here's Nick on biophilia, from an email she sent me sometime after we'd returned from Borneo:

"One of the exercises I do with children is to get them to sit still in an outside space and just pay attention to what their eyes and their ears tell them. They grumble, they muck about and it takes full-on policing to get them to do it. But the results are dramatic. It *always* works. Kids report feeling calmer, feeling happier. Even after just ten minutes of being under a tree, on the grass, by a river, whatever — they come back smiling, surprised, and looking like they've been told a magic word or a big secret. And it's nearly always the start of something good creatively. Biophilia is like the genetic biological clock, so deeply hardwired that it can be revived from the blackened ashes like a phoenix on speed."

99

Fired with Joy

For many years it was my custom at Wimbledon to keep a bird list. This was when I was covering the tennis for *The Times*. It began as a bit of banter with *The Guardian's* then tennis correspondent, Stephen Bierley. Steve is a far better birder than me, not that that's any great mastery. He's more your October Man, and has a British list beyond the 400, the sound barrier in birding.

So yes, there was an element of sporting spice in our contest: "Lesser black-backed flying over Centre Court." "Don't need it. Greenfinch singing by Gate 4." "Bastard!" And so on. Kept us cheerful during a demanding, if often exhilarating, fortnight. It also made sure that we kept looking at birds: a commonplace bird you may not bother to raise your head for, might be the very one that gives most pain to *The Guardian* – and of course, birds do rather tend to bring their own rewards: a screaming party of swifts overflying the pressbox is the sort of thing that keeps you going. But inevitably, the Wimbledon Bird Race was almost all about brief glimpses of very ordinary birds, mostly going from A to B at a rapid rate.

Steve retired from *The Guardian* and moved to Ireland, but I kept the list going because it had become part of the weekly wildlife column I used to write for *The Times*. The annual Wimbledon bird list was part joke, part celebration of biodiversity, part celebration of the fact that wildlife exists even in a city, part acknowledgement of the way that a nice bit of wildlife always makes your day that little bit better.

Shortly before Wimbledon 2014 I was fired from *The Times*. This was unexpected: boxers always say it's the punch you didn't see that does the real damage. I was chief sportswriter, wildlife columnist and, it seemed, part of the fabric of the paper. But no longer. So it was now my job to look resolutely at the future. All the same, it was a fairly difficult transition to make.

I decided that moping at home was the wrong move, so I worked right to the end of my period of notice, and that included one last Wimbledon. I wanted to show the world, future employers and myself that I was not even remotely finished, that I was still full of energy and busting with the ability to tell a good tale. So I kept a last bird list as well. Naturally.

I was walking from Court One back to the press-room when I turned my head to make a routine survey of such treetops as were visible and the sky above. Would I have done it had there not been the list to consider? Probably: I'm always tuned in to such wild stuff as happens to be about. But the list made it more certain that I would raise my eyes and my mind and my hopes. I was looking across Court 18, famous for that match in 2010 in which John Isner and Nicolas Mahut played each other to a standstill over more than 11 hours of tennis, the final set ending 70-68. A court of history, then – and as I looked I had the finest birding moment that Wimbledon had ever yielded.

It was just above the block of flats that lies beyond court 18. Kestrels. Two of them. They must have been young birds, they might even have hatched out on the roof of those flats – and they were playing with the joyous abandonment of the young. They were hovering, and they were stooping at each other; they were like puppies in a game of rough and tumble, or like the fubsy cubs I had seen on the Luangwa beach, except that they were 70 feet up in the air, swooping, curvetting, stall-turning and barrel-rolling: a great exuberant detonation of youth and joy and life.

It'd be overdoing it to say that the professional cares that had been oppressing me vanished at a stroke, but all the same it was a fairly unambiguous message that the world is wide and that to define yourself by your job is somewhat self-limiting. I would love to be able to say that this was an epiphany that allowed me ever after to find a paradise island everywhere; that even in the heart of SW19, even at a time not without bitterness and disappointment, I found myself capable of living a life of unalloyed joy thanks to the wild world and the creatures that inhabit it.

But it was a sign, was it not? The word auspicious means, by etymology, bird-watching. Soothsayers looked to understand the world and what was to happen by reading the signs created by the birds: they called it ornithomancy, divination by means of birds. So I looked at the kestrels and did so not without joy, in the knowledge that the new adventures I was unwillingly setting out on would not be without joy and meaning.

And it was also a confirmation that wildlife and the conservation of wildlife matters to me on a marrow-deep basis. I've often been asked how come I write about both sport and wildlife, what's the connection? Well, the connection is me. Writing about sport is a great way of writing about humans, much more fun than writing about politics or business, for in this sphere the people you write about are involved in perpetual cover-ups. An athlete in competition is emotionally stark naked before you and that allows a sportswriter to write with something of a novelist's freedom.

This has become a bit of a stock-response from me. After all, you have to find a way of answering questions that come at you on a serial basis. Such questions force you to find a formula that is not only plausible but which has the added virtue of actually being true. So I say that I write about wildlife as well as sport, not because sport is not enough, but because humanity is not enough. And then I go on to add: if sport was abolished tomorrow, I expect I'd find some other way of writing about human beings. But if wildlife was abolished tomorrow – an infinitely greater possibility – then I wouldn't want to live.

100

Pillars of Wisdom

Three pillars. I'll always remember them. Each one at least nine or ten feet high. They gave off a dull shine in the Luangwa sun, for their skin was all metal. Wire in fact: tough, thin strands of wire, perhaps a dozen to the inch. It might have been a modernist work in an art gallery, circa 1922. Or perhaps it was a piece of conceptual art, with the clue in the title: *Pain*. Or *Help*. Or perhaps it was a joke within a joke: a label that read *Untitled*, or perhaps *Wire Pillars*. They guarded the entrance to a long, low building: a strong statement of some kind, though one hard to read at first sight.

And then I got it.

Snares.

Each one of these circlets of wire was a snare. Each one had been captured by patrols of scouts, sometimes removed from an animal. There were maybe 1,000 snares in each pillar. I was on my way to Mfuwe airport to fly home, but I called in at the South Luangwa Conservation Society, SLCS. This is a non-government organisation funded mainly by the tourist lodges; its function is to back up the statutory organisation, the Zambia Wildlife Authority, ZAWA, which is chronically underfunded and undermanned. SLCS see their main job as reducing human-wildlife conflict. There are plenty of humans about in the Valley these days, because there are plenty of jobs and plenty of money to be made from tourists. The tourists come for the wildlife, for the unspoiled bush: so not spoiling it is critical on any number of levels. SLCS had freed 14 lions from snares in the past two years. Dogs, being so mobile, so alert and so curious, have also been found snared more often than their numbers would indicate. Even elephants have been caught and, though they tend to uproot the snare as they go, the wire stays in the leg and creates a permanent and potentially lethal wound unless the snare can be removed.

SLCS scouts are constantly on patrol, with radio manned 24 hours. They dart snared animals, remove the snare, tend to their hurts, and free them. They also look for poachers: here their approach is diplomatic and conciliatory. Elias Mutamba, one of the scouts, told me that SLCS employs 50 scouts, including six women. They are also involved in protecting humans from animals: chilli-fencing, which is a technique of keeping elephants away from crops by putting up fences of ropes laced with chilli, and frightening off crop-raiding elephants with paint-ball guns loaded with chilli juice. They are also involved in education and community work.

The poaching tends to be about cash, not subsistence: bushmeat is worth serious money in Lusaka. And the market for illegal ivory is resurgent: recently they had found a man with a pair of 60-pound tusks, and another with leopard-skin.

The Valley had made me drunk with joy for a week. But that doesn't mean I don't have a sober view of how it operates and how its future is not something you can take for granted. One of the joys of being in the bush is stepping away from the human world, but such pleasures are made possible in our troubled century by the existence of good humans doing a hard and dangerous job and doing it damn well.

101

Storm

Innocent was knowingly using the spotlight and using it well. He found us a bushy-tailed mongoose, a civet, several white-tailed mongeese – my preferred plural – and a few genets. I can never see a genet without wanting to take one home: just curl up in the bag, little beast, and we'll saunter through immigration and all will be fine. They're not cats, for all that they're spotty, pointy-eared, bouncy in gait and usually found in trees. They're Viverrids, which is a sort of catch-all group of carnivores. Gid had a rescued genet that she looked after for a year or two, a beautiful animal she called Pippin. He was full of charm though questionably tame, and in the end he went back to the bush when he reached maturity, no doubt in the hope of an equally beautiful female genet. I sometimes feel that life would be complete if I had the daily company of a genet.

Then we heard a leopard calling across a mahogany glade, the one in which Norman is buried. One leopard? No, surely there were two. The pukus called shrill whistles of distress, the impalas barked savagely. But then came a full leopard saw-blade roar; when the wind is southerly I know a leopard from a hand-saw. Leopards don't roar when they're hunting; it would be counter-productive. Something else was up. Something was going on. Then the two of them emerged from the bush, the female all slinky and sinuous and flirty: couldn't keep her paws off him, nor *vice versa*. The male was all set to oblige her there and then when something disturbed them. They slinked together, shoulder to shoulder into cover – but not before the female had slid herself beneath the male's chin and treated him to a good old rubbing lovey-up (a bit like a cat in the painting of Eden) using the whole length of her slinky body. This was not just a biology lesson in action: this was the pursuit of joy, nothing less. Can any human watching have failed to feel empathy? Could any lone wanderer have failed to feel a pang for the deeper pleasures of life at home?

I felt very content as Abe drove us back to Kapani, though a mite cold, even though I was wearing what counts as cold-weather gear in the Valley. The wind was getting up.

Yes, the wind was getting up, blowing in little gusts and uncertainties, working itself up into a temper. Blinking the dust from my eyes, I decided that the bar represented a strong-point and it would be best to use that as a base. The wind greeted this decision by getting serious, and all the lights went out with an appropriate sense of drama. As I drank my Mosi by Braille, thatch began to blow from the roof and the table-settings began to scoot to the floor. The lights came back on, though uncertainly.

The rain hit us during the main course. Coming at us horizontally, underneath the thatch. It didn't feel like a joyous thing at the time: it was cold, vicious, aiming at your eyes and quite astonishingly wet. The half-dozen of us gobbled our excellent food and retreated to the bar, for shelter rather than refreshment. There we had one of those joyous, laughing conversations that small emergencies encourage. We took pudding – a first-class crème caramel – squatting behind the bar as the rain came in like a series of whips and the thatch blew in our faces. Then there was something of a lull; I used it to take a large whisky to my hut. The storm got up again as I sat in bed and sipped, making a showy percussion against the thatch.

And it remade the world. Again.

Abe and I splashed out in the morning, in the rainy or green or emerald season when the world is mud-luscious and puddle-wonderful. It was as if the first shock of the newness of the rains was over; now it was time to get serious. Time to be seriously wet; time for the Luangwa Valley to get on with the job of turning into a completely different place. The transition period was over. Above the bridge a familiar scream: European swifts in great numbers, not there yesterday, riding in on the front that brought us this latest rain. The swifts that live locally – little, white-rumped and horus, all with white bums, and all of which skim past so fast you can't really tell one species from another – build their nests under the bridge and are always hanging around the place, so when the Euros jetted in on their swept-back wings, they hooked up with the locals out of family solidarity. They were all dancing the spinning dance of happy swifts in a time of plenty: screaming in delight at each other's company and hawking for insects – curves and curves and curves – never a straight line even for a second, as if they had learned this lesson from the river beneath them. I wondered what the sky would look like if each swift left a trail in the sky, the sort of thing that happens when you write your name with a sparkler. It would be a great pattern of circles, ellipses, curves: aerial calligraphy with S and O and C, and all kinds of strange swashes and decorative pen-work.

The road we took yesterday was now impassable and would remain so for the next six months. A flock of Lilian's lovebirds took off like an upwards shower of green sparks; from a livid green plain leapt these still greener birds. But then everything was a shade greener than it had been yesterday: it was as if we had shifted two or three squares along in the wild world's paint-matching chart. It also seemed that everything we saw was a shade more content. I picked out African broadbill on call – actually a noise produced mechanically, by vibration of the wings. Abe as ever, politely waited for me to have first crack; it was the first time I had heard this bird in the Valley, and the first I had seen since a particularly demented adventure with Bob into the Northwest Province. In the generously spreading arms of a baobab, a frenzy of nest-building. The supportive, cupped-hand shapes of baobab branches are deeply attractive to some of the Valley's better architects. There were red-billed buffalo weavers, dark little birds with bright red beaks that build apartment blocks: rambling mansions that contain a dozen or more pairs, each in a separate room. Lesser-masked weavers put together their intricate structures, dapper yellow birds, and with them and still more gorgeous, the red-headed weavers made their own nests, elaborate spheres of beautifully woven grasses and twigs, the birds hanging beneath them like blue tits on a feeder, but instead of filling themselves with peanuts they were weaving these little masterpieces of the nest-builder's art.

There was an elephant lying flat out on the ground, a big male stretched out on his side. Common behaviour at this time of year, Abe said: the elephants get tired from their nights of mango-scrumping on the far banks of the river. Another elephant stood beside him: a friend, we would say if we weren't so terrified of sounding anthropomorphic. My horses do exactly the same thing: one will stand guard in the field when another selects the ultimately vulnerable option of a crashed-out snooze. It was a scene of easy intimacy, of generosity, of favour, of service.

And then, as we passed Chipela Chandombo, a lagoon that never dies, I heard a willow warbler. It sang, boldly and yet questioningly, paused while you might count to 20, and then sang once again. And again, a sweet voice carrying across the still, clear first morning of the world. It was a sound I last heard on the marsh at home; it is the one song whose lisping cadences mean that spring is no longer a promise but fully realised; it is the song of my enlightenment. Now it sang the same song 5,000 miles further south and this time it was all about the joys of the rains: the wonder of this newly-bathed country, now fully realised, refreshed, renewed and remade. It was one of those Luangwa moments, when near and far, home and adventure, myself and the rest of life, and for that matter, heaven and earth, time and eternity all seemed very close – close enough to touch, like the otter in the dyke, like the lions sleeping it off after their latest feast. The passage of the seasons and the cascading events of my own life paused for a half-second or an eternity in the silence between the song and its reiteration.

Acknowledgements

I owe grateful thanks to many people in and round the Luangwa Valley in Zambia, especially

Abraham
Aubrey
Chris
Gid and Adrian
Iain
Innocent
Isaac
Jess and Ade
Jo and Robin
John and Carol
Manny
Nick
Shaddy
Willie

And also to the memory of

Bob
Leah
Norman
Perry